The Revolution in Energy Technology

The Revolution in Energy Technology

Innovation and the Economics of the Solar
Photovoltaic Industry

Xue Han

*Postdoctoral Fellow, Faculty of Business Administration,
Lakehead University, Canada and Science Policy and
Innovation Network*

Jorge Niosi

*Professor Emeritus, Department of Management and
Technology, Université du Québec à Montréal, Canada*

 Edward **Elgar**
PUBLISHING

Cheltenham, UK • Northampton, MA, USA

Published by
Edward Elgar Publishing Limited
The Lypiatts
15 Lansdown Road
Cheltenham
Glos GL50 2JA
UK

Edward Elgar Publishing, Inc.
William Pratt House
9 Dewey Court
Northampton
Massachusetts 01060
USA

A catalogue record for this book
is available from the British Library

Library of Congress Control Number: 2018946004

This book is available electronically in the **Elgar**online
Economics subject collection
DOI 10.4337/9781788115667

ISBN 978 1 78811 565 0 (cased)
ISBN 978 1 78811 566 7 (eBook)

Typeset by Servis Filmsetting Ltd, Stockport, Cheshire
Printed and bound by CPI Group (UK) Ltd, Croydon, CR0 4YY

Contents

Contents

1. Introduction

1.1 THE ORIGIN OF THE STUDY

Since the late 1970s the Chinese economy has been fast catching up in many industries, including information and communication technologies, vehicles, and other sectors. The solar photovoltaic industry is different from the other industries. It is the only sector in which the Chinese have claimed a large market share in the world economy. How was this accomplished? What are the factors leading to the success of the Chinese solar PV companies?

In the study of the solar PV sector we found that Chinese domestic factors alone cannot explain their success. The sector needs to be studied in a systematic way and at the global level, because its dynamics are global.

Therefore the topics we decided to study in the solar PV sector with the theories of sectoral system of innovation are as follows:

- How is the sector evolving in both technological and economic perspectives? Does the evolution help the global diffusion of the technologies? How does the technological diffusion promote the development of the sector in developing countries like China?
- What are the respective functions of the innovation-active components in the sectoral system of innovation, including scientists and small businesses? What are their innovation performances and their contributions to the development of the sector?
- Is there a developmental imbalance in the different regions of the world? And if this is the case, what are the factors leading to those imbalances?

All these questions are to be explored in the theoretical framework of sectoral system of innovation.

1.2 STARTING POINT

In order to explore the new understanding of the industry and maintain the academic value of the study, several aspects have been reviewed. We found that the solar PV sector has not been extensively studied:

- When searching the keywords 'sectoral system of innovation' and 'solar photovoltaic', we found just four publications in Scopus. Only one of them is really about reviewing the solar photovoltaic sector with the theories and methodologies of the sectoral system of innovation, and the case study is about just three countries. When the keywords 'sectoral system of innovation' and 'solar photovoltaic' are used to search Google Scholar, there is no other paper on the same subject. A few papers have been written on the global development of the sector so far.
- From the standpoint of 'clusters': when searching the keywords 'clusters' and 'solar photovoltaic', there are 19 publications in Scopus and just four papers are directly related. One is about California, another is about Norway, one is about China, and the last is about Taiwan. There is no complete study of the clusters of solar PV sector in the world.
- From the standpoint of 'star scientist': nothing can be found when searching for the words 'star scientist' and 'solar photovoltaic' in the Scopus database. When the combination of the two phrases is searched in Google Scholar, there are no papers about the same subject.

After the complete literature review was done, we found that these important aspects of the industries have not yet been studied. This virgin and fast-growing sector is waiting to be explored to formulate new findings for similar high-tech industries. Perhaps the sector is too new for economists and management scientists alike to have invested major efforts in its understanding *as a sector*.

1.3 THEORETICAL FRAMEWORK

Our theoretical background is based on evolutionary economics, one that includes non-linearity and inflexion points. We argue that the solar technological system has been improving its performance over the decades since the late 1950s, and recently crossed a landmark point where the performance of the technical system is now speeding up, as seen in Figure 1.1.

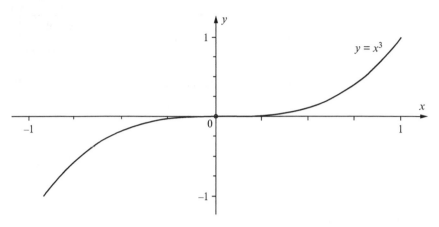

Note: In this figure the x variable indicates time and y indicates performance (cost reductions).

Figure 1.1   Non-linearity in the performance of the solar energy system

1.3.1   The Sectoral System of Innovation

In order to formulate an integrated view of the main dimensions of the sectors and what may account for the differences across sectors, the sectoral system of innovation (SSI) concept is useful. It was put forward by Franco Malerba (2002; 2004; Malerba et al., 1999), according to whom 'A sectoral system is a set of products and a set of agents carrying out market and non market interactions for the creation, production and sale of those products' (Malerba, 2002: 247). The SSI approach highlights a particular set of points: knowledge and structure are key elements; the role of non-firm organizations such as universities, government organizations, local authorities and institutions, and rules of the game such as standards, regulations, labour markets; the dynamics and transformation of sectoral systems are also emphasized.

According to Niosi and Zhegu (2010), the SSI approach emerging from the work of Malerba is as potentially fertile as the previous components of the innovation system perspective. The SSI addition sheds new light on the complexity of the innovation process and helps to understand the trajectories such as how sectoral systems interact with national and regional systems, how sectoral policies are to be understood in the light of national policies and why some countries pull ahead while others fall behind.

The sectoral system of innovation includes the following components:

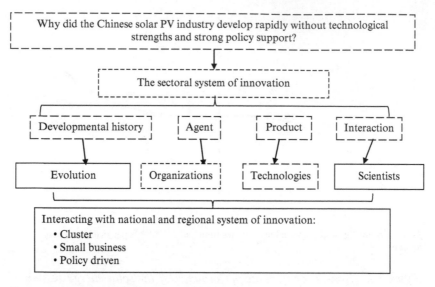

Figure 1.2   The study reasoning route

- Different agents: large firms, small firms, public research organizations, universities and governments.
- Technologies and innovations: the categories of innovations, the process of the innovation produced, the interaction of the organization and technology evolution.
- The institutions: factors including organizations, standards, regulations, labour markets all influence the whole system.

These components will be integrated first by exploring the evolution of the sector; the outstanding results about the different components in the evolution are explored further in the chapters to follow (see Figure 1.2).

1.3.2   The Evolution of the Sector

According to the literature review produced by Malerba (2007), there are two basic models to study sector evolution: sector life cycle models (Industry Life Cycle: ILC), based on the product life cycle (PLC) on one side, and history-friendly models on the other. Since the late 1970s, several studies using the PLC–ILC model have highlighted the fact that a large number of industries follow a life cycle in which a radical innovation and the related entry of new producers that introduce new products is followed by demand growth, a greater emphasis on process innovations and a selection process which ultimately leads to a concentrated market structure and the decline of innovation (Abernathy and Utterback, 1978; Utterback, 1994). But it has been convincingly shown that the dynamic sequences differ from one sector to another (Klepper, 1997; Geroski, 2003; Malerba, 2007). Thus, individual-sector case studies are necessary to see the real industrial dynamics, particularly in high-technology sectors such as biotechnology, information technologies, nanotechnology and solar photovoltaic; these sectors became prominent after the PLC had adopted its canonical form in the 1960s and 1970s. In the meantime, some case studies have been developed using history-friendly models, for example in the computer sector (Malerba et al., 1999; 2001), the pharmaceutical sector (Malerba and Orsenigo, 2002), as well as in other industries such as software and chemicals.

In this research quantitative analysis and the cases studied will be integrated to explore the evolution paths of the solar PV sector.

1.3.3 Star Scientists

When any high-tech sector is studied, the contributions of the scientists need to be explored. Since Edith Penrose (1959) wrote the first scholarly work suggesting that the growth of firms depended on their human resources, highly qualified managers and industrial scientists have been in short supply and, in addition, existing companies are usually employing them. Those companies that are able to hire and retain this qualified human capital will have a sustained advantage over those who are not.

On the basis of Penrose's work, several successive lines of theoretical thought and empirical work appeared in the human resources and strategy fields, linking the competencies of the firm to its performance. The resource-based theory of the firm developed to argue that highly performing firms based their advantage on a series of internal resources, among which human capital played a prominent role (Barney, 1991). Sustained competitive advantage and the related sustained performance come from resources that 'a firm controls that are valuable, rare, imperfectly imitable and not substitutable' (Barney et al., 2001, p.625). These resources are composed of managerial, but also organizational and informational elements.

A second line of thought came with the competence view of the firm. For these authors, resources are valuable only if they translate into competencies: the capacity to successfully combine those resources, incorporate new technical and scientific knowledge, and to attract venture and intellectual human capital, be it administrative, scientific or other. Resources are important if and only if they can be organized in such a way that they deliver performance (Hamel and Heene, 1994). Following this approach, Colombo and Grilli (2005; 2010) argued that the competencies of the founders are key in new-technology-based firms. When they speak about competencies, they are referring to technical work experience; however, they found that new-technology-based firms have superior performance when the team of founders includes people with both economic-managerial and scientific and technical education. In addition, skilled human capital is able to search for new knowledge and new competencies.

In sum, many empirical works have confirmed the link between managerial talent, including scientific and technical, and the long-term performance of the firm, particularly the high-technology-based firm (Colombo and Grilli, 2005; Hitt et al., 2001). Also, advanced human capital is linked to innovation, attraction of venture capital and growth in a positive feedback loop.

As founders with strong backgrounds in science and technology became more and more important in the high-tech firm's development, scientists with the spirit of entrepreneurship have drawn the attention of researchers. Who will contribute more to the development of firms and sectors? How do we recognize these scientists? What are their ways of connecting academic research with business entrepreneurship? What is the performance of their academic entrepreneurship? All the above questions need to be answered.

But not all scientists can contribute to the development of the sector. Lynn Zucker and her colleagues at the University of California Los Angeles (UCLA) launched a small but influential addition to this line of thought. They argued that the biotechnology revolution was the work of star scientists, those biochemists, biologists, medical doctors and other scientists who had published a large number of articles and appeared as the inventors of several influential patents (Zucker et al., 1994; Zucker and Darby, 1996). These stars were often the founders and advisors of biotech companies.

In terms of the ways in which star scientists can contribute to the development of the firms, it is necessary to examine the role of the star scientist in the technology transfer from universities and institutes to the industries. Some of these roles include licensing their patents, establishing university spin-offs (USO), getting listed on the board of directors of start-ups, acting as chief scientists and so on. As to the factors explaining the growth of these spin-offs, using a database of 149 university spin-off companies, Walter et al. (2006) argued that network capabilities and entrepreneurial orientation are key variables explaining the performance of these USOs. Other authors have found that spin-offs from different US universities have very different levels of performance. More entrepreneurial universities have a much better score as licensors of technology to academic spin-offs. Using a very large sample of US academic spin-offs, Powers and McDougall (2005) found that universities with experienced (older) technology transfer offices (TTO) incubate more successful spin-offs. More productive faculty (in terms of articles and citations) are also involved in more successful spin-offs. Early collaboration with the sector is also linked to spin-off growth.

Some studies show that a large percentage of academic spin-offs are related to the biotechnology and health sciences. Mowery et al. (2001) calculate that some 75 per cent of the patenting and licensing in three of the most research-active universities in the United States (California, Columbia and Stanford) occurred in biomedical research, particularly in biotechnology. The second most important sector they highlight is computer software. Similarly, in the annual survey of intellectual property generated in Canadian universities (Statistics Canada, annual), health sciences appear as number one, although they are not as prominent as in the USA. None of the studies mentioned the academic entrepreneurship of the solar PV sector. So we will focus on the star scientists and their academic entrepreneurships in our specific sector.

1.3.4 Regional Systems of Innovation: Clusters

'Clusters' were defined by Michael Porter as 'geographic concentrations of interconnected companies, specialized suppliers, service providers, firms in related industries, and associated institutions (for example, universities, standards agencies and trade associations) in particular fields that compete but also cooperate' (Porter, 1998: 197–8). A clear condition for the existence of a cluster was the presence of linkages between companies and institutions. Niosi and Zhegu (2010) concluded that external economies, regional knowledge spillovers, cluster absorptive capacity and the existence of anchor tenants are among the reasons why clusters are established. The failure of some clusters while other clusters thrive has been well studied, and the resilience of some clusters is understood.

Like most other high-tech sectors, the solar PV sector has also experienced a marked geographic agglomeration. Vidican et al. (2009) found that the solar photovoltaic sector in the US has been concentrated in the two states of California and Massachusetts, with California hosting the largest share of companies over the years. Mathews et al. (2011) found that in Taiwan, the Fast-Follower Strategy (FFS), a strategy which aims at spanning as many steps in the value chain as possible and as quickly as possible, was adopted to promote the solar PV sector by capturing agglomeration and cluster effects for solar PV technology.

The development of the solar PV sector shows a financial imbalance in recent years. With the withdrawal of government subsidies, several big European companies including Siemens closed their operations in 2013. According to Greentech Media, 112 solar energy companies in the United States and the European Union have declared bankruptcy, closed their doors or have been acquired by competitors under suboptimal conditions since 2009.[1] But at the same time, the solar PV sector grew well in China and Japan. According to Businessinsider.com, China and Japan were the top two countries with the biggest added capacities in 2016.[2]

In order to examine the distribution of clusters in the world and to explore the imbalance of the different clusters, the theories on cluster innovation are employed.

It is to be noted that, because of the EU economic crisis, European countries drastically reduced their additional solar PV facilities in 2016, while China and the United States have taken the lead. India has implemented a plan to produce 20 GW of solar power in 2020. In Latin America Chile (700 MW in 2016) has taken the lead, as has South Africa in that continent.

1.4   METHODOLOGY

1.4.1   Research Routes

After selecting the solar PV sector, the secondary information of the sector was explored and the research questions defined. We searched the academic databases and reviewed the related literature, and only then were our hypotheses set up. The methodology and the databases were selected to explore the findings, and then the conclusions were put forward. By extending the findings to the related areas, policy implications were made. The research route is seen in Figure 1.3.

1.4.2   Data Collection

In order to obtain a complete analysis of solar PV innovation all over the world, data on both patents and publications were employed. In addition, case study and secondary data collection were used for the analysis at the different levels to explore the reasons behind the significance.

1.4.2.1   Patents
Patents are good indicators for assessing technological capability. The business literature argues that the number of patents is an appropriate

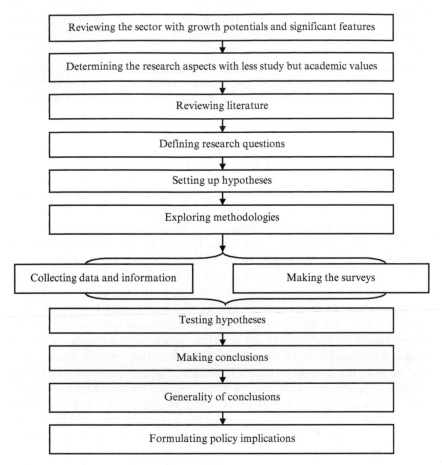

Figure 1.3   Research route of the study

indicator for comparing the innovation performances of companies in terms of new technologies, processes and products (Cassiman et al., 2008; Gittelman, 2008). Even the strongest critics of the general use of patents as performance indicators (Arundel and Kabla, 1998; Mansfield, 1986) admit that patents could represent appropriate indicators in many high-technology sectors. Consequently, the identification of the inventors listed in patents provides key information on the history of R&D processes related to a technical invention and thus a means for retracing knowledge flows through innovation systems or regional clusters of firms. Also, a growing number of researchers use patent citations as indicators of the R&D output of firms, or as determinants of innovation performance that could impact on their growth. Unlike a simple counting of patents, which is purely quantitative, patent citations also allow a measurement of patent quality because there appears to be a positive relation between a patent's importance and the number of times that it is cited. Patent citations can be very useful as indicators of a patent quality in economic studies of biotechnology-firm innovation and performance (Jaffe and Trajtenberg, 2002).

Hu and Jaffe (2003) initiated a new line of work examining patterns of knowledge diffusion from advanced countries to latecomer catch-up countries with their study of United States Patent and Trademark Office (USPTO) patents taken out by Korea and Taiwan over the 22-year period from 1977 to 1999. Four stylized facts emerged from their work which have formed a benchmark for subsequent studies of knowledge diffusion, or what can be called knowledge leverage by latecomers.

Since then further work has been devoted to taking the analysis to the industry level. Starting with the DRAM (Dynamic Random Access Memory) industry, Lee and Yoon (2010) investigated patterns of catch-up by Taiwan and Korea, and then they extended the net to include China. Analysing the patents from the USPTO by Korea and Taiwan over the period 1985–1999, Lee and Yoon (2010) argued that they had found evidence that with regard to relative citation propensity, the order of patent citation follows the order of national entry into the industry, namely that Japanese firms tended to cite US patents; Korean firms tended to cite Japanese patents; and Taiwanese firms tended to cite Korean patents. Lee and Wang (2010) then extended these results to China, arguing that Chinese firms tended to cite Taiwanese patents, and that as the latecomer, China exhibited the lowest level of intra-national knowledge flows (reflecting low absorptive capacity).

The next industry so studied was flat panel displays (FPD); Hu (2008) used US patents registered by the top five Taiwanese FPD manufacturers to trace their knowledge sources of FPD technologies. The finding suggests that the knowledge source in latecomers, such as Taiwan, is mostly secured from Japan on specific core technologies, rather than from the US. Jang et al. (2009) further assessed the innovative capability and international knowledge flows amongst technological forerunners (US and Japan) and latecomers (Taiwan and Korea) in the FPD sector, and confirmed that the latecomers (Korea and Taiwan) leveraged significant knowledge flows from the technological leaders (US and Japan). But in contrast with earlier studies, Jang et al. (2009) had found that Japan dominates knowledge flows for Korean firms in the FPD industry (Japan accounting for 56 per cent of total citations by Korean firms at the USPTO between 1976 and 2005, compared with only 20 per cent for the US); likewise Taiwan firms' patenting favoured Japanese patents. Japan has used the US as a knowledge source, but with less divergence (39 per cent for Japan vs. 34 per cent for the US). This too presents a very interesting finding that calls for further examination in emerging industries such as solar PV.

Lee and Jin (2012) then turned their attention to the mobile telephone industry, covering patents taken out at the USPTO by Korean, Taiwanese and Chinese firms over the period 1976–2008, and found again that in terms of relative citation propensity, the order of patent citation follows the order of entry into the industry, with Japan following the US, Korea following Japan, Taiwan following Korea, and China following Taiwan.

In the study, the patent numbers are counted as proof of the innovation capabilities. As the focus is on the solar PV sector, the keywords selected are 'solar cell', 'solar cells', 'photovoltaic cell' and 'photovoltaic cells'. After studying the patent data in the USPTO and in the European Patent Office (EPO), we found that the data in the USPTO is more applicable than that in the EPO for the following reasons:

1. There are many more patents in the EPO and Chinese Patent Bureau database than in the USPTO, but the above keywords in the EPO and Chinese Patent Bureau database do not produce results that are as exact as in the USPTO, which means that by reviewing the patents randomly selected from the EPO database, some with the keywords in the abstract are not in the solar PV domain. And when the patents issued in the USPTO in the databank of the EPO are compared with the patents directly from the USPTO, there is a big difference in terms of quantity and quality. In order to find more comprehensive data to analyse the sector, the USPTO database is selected for the patent analysis.
2. The list of issues within each individual patent in the USPTO is more comprehensive than that in the EPO. For example, only the patents in the USPTO include the inventor locations.
3. As the most prolific inventing country, the United States assignees own nearly 50 per cent of the solar PV patents. In addition, competitors in Japan, Germany, Taiwan or the People's Republic of China also patent their inventions in the United States in order to protect them from potential infringers.

Therefore the patent data in the USPTO has been selected as the basis for our patent analysis. Apart from the chapters covering the introduction, characteristics of the sector and conclusion, the different issues in the individual patents are taken as samples in the other five chapters (Table 1.1).

The revolution in energy technology

Table 1.1   The dimensions of patents employed in the different chapters

Issues	Chapter 2	Chapter 3	Chapter 4	Chapter 5	Chapter 6
Abstract	*	*		*	*
Year issued	*	*		*	
Assignee's name		*		*	*
Assignee's state				*	*
Assignee country		*	*	*	*
Inventor's name				*	*
Inventor's city		*		*	*
Inventor's affiliation		*	*	*	*
Inventor's state		*		*	*
Inventor's country		*	*	*	*

1.4.2.2   Publications

After reviewing several publication databases, we selected *Scopus* to analyse the solar PV sector due to its quality in terms of journal selection and better taxonomy for academic research papers. All the papers in the databases were searched with the keywords 'solar cell', 'solar cells', 'photovoltaic cell' and 'photovoltaic cells', and then all the publications obtained from these keywords were taken as the new database for analysis. As it was found that nearly all the authors have publications in the other domains, their publications in the other domains were also searched and analysed to explore their academic behaviour and knowledge transfer.

In order to see the relationship between the academic behaviour and entrepreneurship for the star scientists, both the patents and publications for the star scientists were studied to see the interrelationship.

1.4.2.3   Case study

In order to explore the answers to how the SSI evolved, the differences among the clusters and why academic entrepreneurship is limited in the solar PV sector, case studies using interviews with some key persons were employed.

1.4.2.4   Secondary data collection

For some key case studies where there was no response to the survey request, secondary data collections were conducted to obtain the relevant information. Secondary data were found in the annual report, statistical report, news published in the companies' websites and data in the government departments' websites. For example, in order to ascertain the influence of venture capital in academic entrepreneurship, annual studies from the venture capital association were employed.

1.5 STRUCTURE OF THE VOLUME

There are nine chapters in the volume and they are organized as follows:

Chapter 1 is the 'Introduction', which begins with the research questions raised after reviewing the solar PV sector. After highlighting the starting points, the theoretical frameworks and the methodology are introduced. Then the complete research route is described and the structure of the volume is included.

Chapter 2 introduces the various key points of the solar PV sector in terms of industrial performance, technologies and regulations. This chapter serves as the foundation for understanding further specific studies.

Chapter 3 is about the innovation cascade of the sector. By comparing Industry Life Cycle (ILC) and Product Life Cycle (PLC) models, the innovation cascade is outlined for the solar photovoltaic sector.

Chapter 4 analyses the catch-up of the Chinese solar PV sector. The new techno-economic paradigm, government support, the human resource context, and integrative production capabilities are extracted to formulate the answers.

Chapter 5 analyses some 4400 US patents on solar photovoltaic (PV) technologies, protecting inventions made in the United States, Japan, Germany, Taiwan, South Korea and other OECD and emerging countries, in order to find out the spatial distribution of inventors. It is found that there are clusters of solar PV sectors in different countries. With the exception of the Silicon Valley, a special cluster with a unique 'bottom up' region in the solar PV sector, and government laboratories anchoring PV clusters in Taiwan, a major multinational corporation anchors all other clusters. The clusters are growing in Asia, resilient in the US, and declining in Europe.

Chapter 6 analyses the contribution of star scientists to the development of solar photovoltaic technology. It is found that the technology has been launched and keeps moving forward with major investments from large user companies and individual efforts from universities and academic stars. In contrast with biotechnology, star scientists, whatever their contribution, are comparatively minor players in solar photovoltaic technology.

Chapter 7 is about the features of the solar PV sector. After studying the evolution of the sector, its geographic agglomeration and the behaviours of the star scientists, some divergence in terms of innovation from the other high-tech industries is detected. By comparing it with the semiconductor

Table 1.2 Description for each chapter

Chapter	Themes	Research questions	Theories	Methods	Theoretical and empirical contributions
1	Introduction				
2	Key points of the sector				
3	Evolution and innovation cascade	1. What is the evolution of the solar PV sector? 2. What are the characteristics of the technological trajectory?	Sectoral system of innovation, Product Life Cycle and Industrial Life Cycle	Quantitative and case studies	Redefinition of term 'innovation cascade', the debate of PLC–ILC theories
4	China catch-up	How can the Chinese solar PV sector complete its catch-up?	Catch-up theories	Case studies and secondary data	Horizontal technological policies, the key entrepreneurs and the integrative production capabilities are the important factors
5	Cluster	1. Are there solar PV innovation clusters? 2. If yes, where do the clusters locate? 3. What are the differences among different clusters?	Cluster	Quantitative analysis and case studies	There are 23 clusters in the world, yet there are imbalances in the cluster development in Europe, North America and Asia
6	Star scientist	1. What is the definition of the star scientists for the solar photovoltaic sector? 2. How much academic entrepreneurship is in the solar PV sector? 3. What are the roles of venture capital, universities and technology transfer offices?	Star scientists	Quantitative analysis, case studies and secondary data	The definition of star scientists; academic entrepreneurship and venture capital are limited; famous universities and their technology offices do not play an important role in the academic entrepreneurship
7	Demand-driven	What are the characteristics of the solar PV sector?	Evolution of the sector	Comparison study	The solar PV sector is a demand-driven sector
8	Grand challenge policy	What is the relationship between innovation cascade and the grand challenge policies?	Grand challenge policy	History study	Grand challenge policy partly explains the innovation cascade
9	Conclusion				

sector, it is found that the solar PV sector is one whose innovation is mostly driven by demand, but so far not pushed by science and technology progress. Three special aspects are put forward. Further study is called for about whether its distinctiveness is significant enough to be a sub-category of the high-tech sector.

Chapter 8 links two recent concepts about innovation: those of grand challenges and innovation cascades. In addition, the evolution of the innovation policy is discussed.

Chapter 9 is the conclusion. The contributions of the studies are high-lighted, the policy implications are made, the limitations of this enquiry are noted, and further research directions are suggested.

The themes, research questions, theories, methodologies, theoretical or empirical contributions of each chapter are outlined in Table 1.2.

NOTES

1. http://dailycaller.com/2014/12/08/112-solar-companies-have-closed-their-doors-in-5-years/.
2. Source: http://www.businessinsider.com/best-solar-power-countries-2016.

2. Some key elements of the solar photovoltaic sector

2.1 SIGNIFICANCE OF THE SECTOR

Today, about 80 per cent of the world's energy production comes from fossil fuel, and to date, coal is the major source of electricity, with a share of 42 per cent of electricity generation and will continue to be the prime source of electricity in many countries in the coming few decades. But after a few years, it may be more difficult to generate electricity from fossil fuels such as coal. Different organizations such as the US Energy Information Administration (EIA), International Atomic Energy Agency (IAEA), International Energy Agency (IEA) and the World Energy Council (WEC) have published their projections of future energy demands for 2020, 2030 and 2050, which show that only the clean energy systems have the capacity to neutralize the environmental impacts of coal.

Clean energy is renewable energy that could have the capability of meeting the energy demands as well as mitigating global warming. In the past few years, it has been observed that renewable energy technology is steadily maturing and its share of energy production has been going up. According to Bloomberg New Energy Finance (BNEF), wind, solar, biomass and waste-to-power, geothermal, small hydro and marine power are altogether estimated to have contributed 9.1 per cent of world electricity generation in 2014, compared to 8.5 per cent in 2013. This would be equivalent to a saving of 1.3 gigatons of CO_2 as a result of the installed capacity of those renewable sources.

Among renewable energy sources, solar energy has its own distinctive advantages: it cannot be monopolized by a handful of countries, as is the case with fossil fuels. It has neither excessive maintenance and management costs nor conversion mechanisms producing troublesome emissions, and it can easily be integrated into both public and private buildings without external environmental impacts, such as those incurred by wind turbines. It is modular, and comparatively inexpensive to install. According to Technology Roadmap: Solar Photovoltaic Energy – 2014 from the International Energy Agency (IEA), solar energy could be the largest source of electricity by 2050. According to Pew Charitable Trusts, in 2013, for the first time, solar outpaced all other clean energy technologies in terms of new generating capacity installed, with an increase of 29 per cent compared with 2012. From then on, the solar PV generation capacity keeps on growing. This is due in part to ongoing price reductions, including significant cuts in manufacturing costs, but also as a result of investment shifting from small-scale projects to less expensive large-scale ones. Added to this is the fact that electricity prices have increased in general. This has led to a situation where grid parity (the moment when electricity from solar panels costs as much as or is even cheaper than electricity purchased from the grid) is within reach. China was the top global market in 2013 with 11.8 GW. Germany topped the European market with 3.3 GW, while the UK was runner-up with 1.5 GW. Europe's role as the PV market leader has come to an end, but various markets within Europe still have almost untapped potential.

The co-research carried out by the United Nations Environment Programme (UNEP) and BNEF in 2014 shows that solar PV is the renewable energy sector with the highest global new investment and with the highest growth rate in 2014.

The costs of solar generation are meanwhile continuing to fall. According to BNEF, the global average levelled cost was $315 per MW/h for crystalline silicon PV projects in the third quarter of 2009, but this had fallen to $129 per MW/h in the first half of 2015, a reduction of 59 per cent in just five and a half years. During the same period of time, the onshore wind cost has remained at around $100 per MW/h, and offshore wind costs have even increased from $150 to more than $200 and then have decreased to around $175. Compared to wind cost, the cost advantage of solar is becoming more and more significant, which shows that the age of solar energy is coming.

2.2 TECHNOLOGICAL INNOVATIONS

2.2.1 Technology Categories

The sun provides 10 000 times the amount of energy actually used by humans every day. Using the totally renewable and sustainable source from the sun, solar technologies have two main types of application: heat and electricity. These two applications can be divided into three categories: (1) photovoltaic (PV), which directly converts light to electricity with the solar cells; (2) concentrating solar power (CSP), which uses heat from the sun (thermal energy) to drive utility-scale electric turbines; (3) heating and cooling systems, which collect thermal energy to provide hot water and air conditioning. In this research, only the solar PV sector, the largest and most promising, is studied.

Solar PV cells are at the core of solar energy technologies. Solar cells made of silicon have rapidly become the key component of solar modules (Parida et al., 2011). These cells are specialized semiconductors that convert solar light into electrical energy, with different levels of efficiency. Other elements are less important, but are gaining a more central role in the efficiency of solar equipment. Solar glass is among them. Initially, conventional glass was used to protect solar panels from damage caused by hail or other falling objects. New advanced glass increases the efficiency of solar PV systems.

This science-based set of technologies of solar PV systems has evolved enormously in its 50-year development,[1] but just around 2008, innovation began to boom. The development of the sector has been quite slow, particularly when compared with biotechnology and information technology. The reason is partly that solar energy has always had strong competition from other sources of energy (coal, gas, oil, hydro and nuclear, as well as wind among renewable energy sources). The technological development of the solar PV sector began to boom since the beginning of the twenty-first century due to the strong policy support beginning in the 1990s from the different central governments in several countries (see Figures 2.1 and 2.2).

It is considered that the science and technological innovation boom from the 2000s on is due to the strong support of governments (see Table 2.1).

Currently, there are three generations of technology in the solar PV sector:

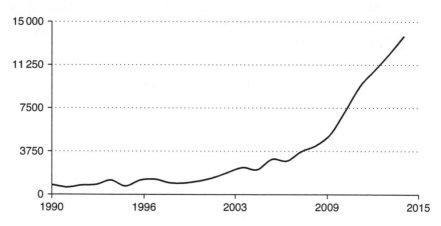

Source: Scopus.

Figure 2.1   Solar cell publications in Scopus (1990–2015)

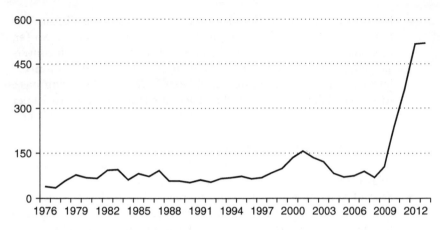

Source: USPTO.

Figure 2.2 Solar cell US patents granted by year (1976–2013)

Table 2.1 Timetable for solar energy policies in the representative countries

Year	Germany	Japan	USA
1991	1000 PV Rooftop programme		
1992			Energy Federal Policy Act
1994		Official launch of New Sunlight Plan	
1998	100 000 PV rooftop programme		1 000 000 PV Rooftop programme
2000	Renewable Energy Sources Act	Green Procurement Law	
2002		Law on new energy for power generation	
2004	Revised Act on Renewable Energy Law		
2005			Revised Energy Law
2008	Newly Revised Act on Renewable Energy Law	PV industry Subsidy policy restarted	
2010			10 000 000 PV rooftop Programme
2011	Third decrease in feed-in tariff (FIT)	Renewable energy Law	
2012		FIT introduced	

- First generation solar PV cells are made in crystalline silicon. The cells are cut from a silicon ingot, casting, or grown ribbon. So far, this generation is dominating the market because of its high conversion efficiency, defined as the percentage of sunlight that is converted into electrical energy, as well as its extensive manufacturing base. Mono-crystalline PV cells today have an efficiency of 16 to 20 per cent, while the cheaper-to-produce multi-crystalline PV cells achieve a slightly lower 14 to 15 per cent efficiency. Crystalline solar PV cells are usually interconnected and encapsulated between a transparent front (typically glass) and insulating back cover material to form a solar PV module, which is usually mounted in an aluminium frame.

- Second generation solar PV cells are referred to as thin film because thin layers of PV materials are deposited on low-cost substrates like glass, stainless steel or plastic. Their advantage is that they are significantly cheaper to produce, but they have much lower efficiency levels. The oldest and most prevalent thin-film cell technology is 'amorphous silicon' with a conversion efficiency of just 6 to 7 per cent, while hybrid 'amorphous/micro-silicon technologies' achieve about 8 per cent. Other thin-film technologies, using compound semiconductors, such as germanium (an amorphous silicon thin-film), cadmium telluride (CdTe) or copper iridium diselenide (CIS), have achieved commercial conversion efficiencies of up to 11 to 12 per cent. These improvements in thin-film efficiencies have led to a very rapid expansion of this segment of solar PV technologies in recent years; their market share has risen from less than 5 per cent in 2004 to over 22 per cent by 2008 (Kirkegaard et al., 2010).

- The third generation of solar PV technologies is not yet being deployed on a large scale. Work to date suggests there is scope for improving solar cell performance by exploring approaches capable of giving efficiencies closer to thermodynamic limits. Low-dimensional structures seem to show some promise due to the small dimensions and new features offered. Common third-generation systems include multi-layer ('tandem') cells made of amorphous silicon or gallium arsenide, while more theoretical developments include frequency conversion, hot-carrier effects and other multiple-carrier ejection techniques. Emerging photovoltaic technologies include: Copper zinc tin sulphide solar cell (CZTS), and derivatives CZTSe and CZTSSe; Dye-sensitized solar cell, also known as 'Grätzel cell'; Organic solar cell; Perovskite solar cell; Polymer solar cell; and quantum dot solar cell.

2.2.2 Measurement of Solar Cell Efficiency

Energy conversion efficiency is taken as the most important indicator to measure the technology progress. According to Green (2009), over the last two decades terrestrial cell measurements have evolved to the stage where independent laboratories measure the same result for standard silicon cells within 1–2 per cent. As a result of early initiatives by SERI (the US Solar Energy Research Institute, now National Renewable Energy Laboratory, NREL) that encouraged the development of highly efficient silicon cells, several key silicon cell results were measured at NREL in the early 1980s, the beginning of what will be referred to as the 'modern phase' of silicon cell development (see Figure 2.3).

2.3 REGULATIONS OF THE SECTOR

Governmental subsidies have played a prominent role in the growth of solar power. According to Sahu (2015), the top ten solar PV power producing countries (USA, Japan, Germany, France, Italy, Spain, China, Australia, Belgium and the Czech Republic) mainly depend upon their policies such as instruments like the feed-in tariff (FIT), net metering, quotas with green certificates, low interest bank loans, renewable portfolio standards (RPSs), countries' national renewable energy targets, Investment Tax Credit (ITC), market premiums, and reverse auctions for the development of solar energy. Without such policies, the high cost of generating solar power would prevent it from competing with electricity from traditional fossil fuel sources in most regions.

But the sector's economics is changing. Over the last two decades, the cost of manufacturing and installing a photovoltaic solar-power system has decreased by about 20 per cent with every doubling of installed capacity as a result of enhanced cell efficiency and the falling cost of solar cell panels. The cost of generating electricity from conventional sources, by contrast, has been rising along with the price of natural gas, which heavily influences electricity prices in regions that have large numbers of gas-fired power plants. As a result, solar power has been creeping toward cost competitiveness in some areas.

According to New Energy Outlook 2016 from BNEF, solar energy costs, which have already fallen by 80 per cent since 2008, will fall another 60 per cent to an average cost of $40/MWh around the world by 2040. The precipitous fall of solar energy costs will lead to the rapid installation of solar, which will account for nearly half of all new capacity installed around the world in the coming decades. One third of this will be on rooftops, and it

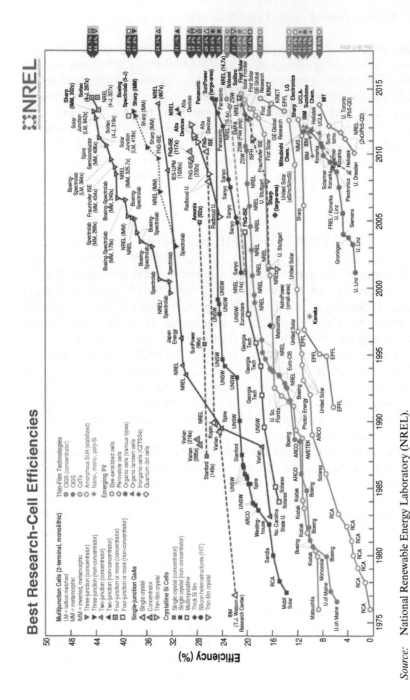

Source: National Renewable Energy Laboratory (NREL).

Figure 2.3 *The evolution of research solar cell efficiency (1975–2014)*

will be accompanied by huge growth in battery storage. By 2040, solar will supply 15 per cent of all electricity demand.

With the cost reduction facilitated by technology development, the sector is becoming more and more challenging and needs to be regulated with new rules. Instead of gradually withdrawing subsidy policies, Mahalingam and Reiner (2016) concluded that, based on a scenario analysis, if the current levels of carbon price were to exist post-2020, both Italy and Spain would find it rather difficult to increase the penetration of renewables in their electricity mix. A high subsidy world, on the other hand, would result in the most favourable outcome, particularly for Spain, although it may incur additional costs in comparison to a high carbon price world.

NOTE

1. Although some people associate solar panels with new-age technology, scientists have actually been working with solar cells for nearly 200 years. The evolution of solar panels has been a slow but worthwhile undertaking.

3.   Sector evolution under innovation cascade

There are two basic models to study industry evolution: the time-honoured industry life cycle models (ILC), based on the product life cycle (PLC) approach, and the more recent and fairly incipient history-friendly models (Malerba, 2007). Since the late 1970s, several studies using the PLC, and then the ILC models, have highlighted the fact that a large number of industries follow a life cycle starting with a radical innovation and the related entry of new producers that introduce new designs. This phase is followed by market growth, a greater emphasis on process innovations and a selection procedure which ultimately leads to a concentrated market structure, the emergence of a dominant design, and the decline of innovation (Abernathy and Utterback, 1978; Utterback, 1994). But it has been convincingly indicated that the dynamic sequences differ from one industry to another (Klepper, 1997; Geroski, 2003; Malerba, 2007). Thus, individual-industry case studies are necessary to see the real economic dynamics, and the differences from one sector to the next, in order to avoid fast generalizations, particularly after the PLC had adopted its canonical form in the 1960s and 1970s. In the meantime, some industry studies have been developed, using history-friendly models, for example the computer industry (Malerba et al., 1999; 2001), the pharmaceutical industry (Malerba and Orsenigo, 2002), as well as for other industries such as software and chemicals.

Also, the development of high-tech industries in the last forty years shows anything but a decline of innovation. Instead what we observe is a rapid succession of radical innovations, what we will call an innovation cascade, here the concept is defined, measures are proposed to give it its empirical contours, and the appropriate theoretical conclusions are made.

Also, this chapter explores the contours of the solar photovoltaic sector, a high-tech one that, like other science-based industries, shows uninterrupted innovation from its modest start in the 1950s to the present day. Like aerospace, biotechnology, ICT and nanotechnology, the solar PV sector does not see innovation stagnate and decline, as argued by the PLC–ILC approaches, but on the contrary, it shows continuous novelty, and branching into new paths as well as the creation of new products for new markets. The institutional environment, the 'general purpose technology' (GPT) character of many of these high-tech sectors, and multiple positive feedback processes, can explain the 'anomaly'. The institutional milieu and the positive feedback effects of the innovation cascades are absent in the PLC–ILC perspective, but is a central element in the innovation systems approach.

3.1 THE EVOLUTION OF INDUSTRIES FROM LIFE CYCLES TO INNOVATION CASCADES

3.1.1 PLC and ILC

In these models, new products trigger innovation. Later on product innovation declines and process innovation takes up the baton. After a certain number of years, both product and process innovation decline, the industry concentrates and it often moves to countries where the labour force is cheaper. This is the story of textiles, garments, furniture and bicycles in the past, and automobiles, electrical equipment and other industries today. In the PLC–ILC theories, innovation is the result of the search activity of one company or a small group of firms, who are manufacturers based in the richest countries, often the United States. Companies produce novelty to integrate it in their product lines. Over time, the market of the innovating country becomes saturated and the innovators start exporting to other affluent markets. Then competitors appear in this second cohort of countries, and the original innovators invest in this second tier of markets in an effort to crush the new competitors or at least slow their progression. Yet, in spite of these efforts the industry becomes more competitive, technology more standardized, and foreign direct investment heads towards third world countries. During this progression, innovation declines, first in products and then in processes.

The industry life cycle approach puts more emphasis on the structure of the industry than on the evolution of the product itself. As in the PLC approach, entry is concentrated in the initial phases of the cycle. The ILC includes a rapid shakeout in the maturity phase, during which the number of firms declines, and the less efficient competitors are eliminated. The emergence of dominant designs makes economies of scale, not innovation, the crucial competitive factor (Agarwal et al., 2002). Similarly, according to other observers, in the mature stages of the life cycle, entry is more related to filling niches than to conducting radical innovation (Agarwal and Audretsch, 2001).

In addition, some authors have added nuances to the PLC–ILC models.

Some of them have noticed that innovation does not necessarily decline in mature industries, when measured by patent activity (McGahan and Silverman, 2001). Those who – in agreement with the PLC–ILC approach – see the number of patents most often falling over the life cycle, oppose this point of view (Haupt et al., 2007). Yet other authors, using a variety of indicators, namely patents and scientific publications, found a variety of situations between invention and innovation. In the science-based industries, the study of the evolution of publication may be extremely useful, as publication either precedes or accompanies the evolution of patents (Järvenpää et al., 2011). Also, Lee and Veloso (2008) found that architectural knowledge dominates innovation during the early phases while component innovation becomes more frequent at the later phases of the life cycle.

A few authors have criticized the PLC–ILC models for their ambiguities about the length and time of the phases, and the eventual coming of dominant designs. The validity of these models in the high-tech industries has been questioned (Grantham, 1997). What exactly declines over time: families of products (i.e. aeroplanes)? Or specific designs, such as turbo-prop aeroplanes?

In their review of the literature, Cao and Folan (2012) added other criticisms: some products have known second lives, particularly under the product renewal and marketing efforts of original innovators. Also, they pointed out that networks of innovators and other inter-company relationships are absent from the PLC–ILC models. Lambkin and Day (1989) had already underlined the fact that the PLC approach ignored the differences between large and small pioneers, those that enter by in-house innovation and those that arrive by acquisition, licensing and joint venture. Also, the approach does not distinguish 'strategic windows of opportunities' for entrants of different size and competitive position. In addition, the models do not predict when the shakeout will occur. Yet the sceptics of these approaches have been few in number and have not been widely cited or followed. In some areas of management, such as marketing, their popularity has not decreased for half a century.

Among the few papers that have tried to devise a role for governments in the product life cycle, Tassey (1991) had argued that governments should adapt their technological infrastructure to the technology life cycles of their main industries. Robertson and Langlois (1995) added that governments should act as facilitators, letting firms adapt to their particular environments. Yet, Klepper, the father of the ILC theory, made it very clear: 'In summary, apart from an occasional influence of the government/military, the evolution of the six products through their formative eras largely conforms with the PLC' (Klepper, 1997).

Thus, in this family of models, governments have an occasional role in the sequence but usually they keep their hands off the product and industry life cycles.

3.1.2 System Dynamics, Complexity and History-Friendly Models

More recently the sectoral innovation system approach that Malerba, Orsenigo and others have pioneered is more attentive to the specific contours of each set of industries or sectors, where particular cost structures, competitive advantages (natural or built), particular policy incentives or market sizes and structures can affect the evolution of the sector.[1] This type of model allows for different sequences in the role of innovators, different types of demand structures and knowledge flows, and different institutional settings. Without obliterating the PLC–ILC approach, these perspectives allow for a variety of sequences and industrial dynamics. These models argue that technological regimes and demand structures have major roles in industry evolution (Garavaglia et al., 2012; Malerba and Orsenigo, 2015).

The relative importance of these currents is easy to grasp: when searched in the Scopus database, the keywords 'history-friendly models' find 18 articles in refereed journals; 'product life cycle' gets 3100 articles.

The general argument that we put forward here is that science-based industries and sectors (SBIS) show a fairly different set of evolutionary patterns compared to more traditional industries. In SBIS, product variation is overwhelming, the branching out of new industries and the rise of new market segments is widespread. Thus, SBIS are more prone to innovation cascades than low and medium-tech industries (McKelvey and Niosi, 2015). In addition, science-based industries are very much supported by governments.

This chapter argues – as Lundvall (1992), Malerba (2004), Nelson (2005), Niosi (2010) and others have done in the past – that institutions play a key role in the development of high-tech industries. This role has specifically been studied in biotechnology (McMillan et al., 2000; Whitley, 2003), in computers and software industries (Mowery, 1996), and nanotechnology (Roco and Bainbridge, 2005; Niosi and Reid, 2007). Governments subsidize these industries in order to improve human health and industrial competitiveness, and aiming at other societal benefits. They do it through public laboratories (think of the National Institutes of Health or the New Renewable Energy Laboratory in the United States), through academic subsidies to R&D, and through multiple incentives to private-sector innovation, including reimbursable and non-reimbursable subsidies, tax credits for R&D, and subsidized tariffs.

Table 3.1   Four stylized cases of industry evolution

		Initial conditions	
		Concentrated	Dispersed
Long-term trend	Increased or sustained concentration	Cases: satellites and space launchers manufacturing (Schumpeter Mark II industries, following Malerba and Orsenigo, 1996)	Cases: pharmaceutical drugs, computer software services (increasing returns industries following Kaldor–David–Arthur)
	Increased or sustained dispersion	Cases: Semiconductors, computer and telecommunications equipment manufacturing (variation-intensive industries following Niosi, 2000 and Saviotti, 1996)	Cases: biotechnology R&D services, professional equipment manufacturing, solar PV equipment (Schumpeter Mark I industries, following Malerba and Orsenigo, 1996)

Source: Based on Niosi (2000).

In addition, increasing variation is a characteristic of several industries, an evolution that Schumpeter Mark I and Schumpeter Mark II sectors as well as PLC–ILC industries have not captured. The following table summarizes the argument that appears as a development based on the Mark I–Mark II dichotomy (see Table 3.1).

3.1.3   The Patterns of Innovation

Innovation is the engine of economic growth. It is thus critical to understand how it proceeds. For several decades, evolutionary theories using the biological model were applied to innovation (Basalla, 1988; Petroski, 1994; McKelvey, 1998): innovation was supposed to proceed in a leisurely way, over the centuries if not the millennia, one step at a time, in an incremental process. Similarly, organizations and institutions evolved slowly from one form to the next (Nelson and Winter, 1982b; Tushman and Romanelli, 2008). Several evolutionary models, as we have seen, have been advanced to explain this change (Malerba, 2006). For most authors, including the authors of this book, evolutionary technological change is the most frequent. Arthur (2009) calls this incremental technical change 'standard engineering'. Bessant et al. (1994) underlined the fact that continuous innovation is sometimes difficult. Yet the vast majority of authors find that evolutionary innovation is ubiquitous. Companies and governments alike abhor disruptive technological change that may devalue their assets and sunk costs, and cannibalize their products (Christensen, 1997).

3.1.3.1 Radical innovation

Radical innovation (already identified by Schumpeter in his 1939 book *Business Cycles*) appeared and it was deemed analogous to biological change, where saltation (Gould and Eldredge, 1977) and short periods of rapid structural change interrupted long periods of stasis and incremental change. Both in biology and management, radical innovation and saltation were difficult to accept. In biology, the neo-Darwinian synthesis wiped out most ideas about saltation. They have slowly come back in biology since the 1950s and 1960s through the work of B. McClintock (Nobel Prize in physiology 1983). The idea was developed and popularized by S.J. Gould and N. Eldridge. How do the markets accept these complex modifications of product and/or process? In the post-war period, radical innovation appeared in Britain in the works of Gibbons and Littler (1979), Rothwell (1980) and others. A few years later, several authors were discussing the multifarious dynamics between radical innovation, organizations and industry structure (Souder, 1983; Achilladelis et al., 1990; Christensen and Bower, 1996) as well as the importance of the necessary infrastructure for radical innovation to be adopted (McIntyre, 1988).

Radical innovation is also labelled 'discontinuous' or 'disruptive innovation'. Radical innovation, when successful, has a much larger effect on firms' profitability, market share and entire industries (Sainio et al., 2012). Key dimensions of radical innovation include technology novelty (clear advances in frontier technology, as in the iPad compared to the iPod), and market novelty (products that address themselves to new markets, or to markets that were served by other products, such as monoclonal antibodies (MABs). Even if it is often the special activity of entrepreneurial firms, it also occurs in large established companies (O'Connor and McDermott, 2004).

In recent decades, innovation seems to be accelerating; new scientific disciplines appear, and new technical branches of technical knowledge multiply. Thus, it cannot be properly depicted as a smooth path, punctuated by occasional changes in direction. Genomics has given rise to pharmacogenomics, metabolomics and lipidomics. It looks much more like a river where fast-flowing water evolves from rapids to waterfalls, and splits into several diverging flows that sometimes merge with other flows to form new estuaries. The concept of innovation cascades circumscribes evolutionary change (Antonelli, 2008; 2009; Berkers and Geels, 2011; Delapierre and Mytelka, 2003; Lane, 2009a; 2012). Rothwell and Wissema

(1986) had suggested that radical innovations arrive in clusters, much in line with the Schumpeterian view of business cycles. This chapter argues that innovation cascades are becoming much more frequent today for several reasons: because of the rise of science-based industries (Pavitt, 1984), the increasing number of research universities in a growing number of emerging countries, more linkages between these loci of knowledge creation, and faster technology diffusion. Fastest imitation also increases the probability of new combinations between different strands of knowledge. Innovation cascades have a definite Schumpeterian flavour.

3.1.3.2   Incremental innovation

Evolutionary or incremental innovation (small, continuous improvements in technology and organization) is the most abundant type of innovation. Its predominance over other forms of innovation is easy to accept. Companies and individuals tinker with what they know best. Such behaviour reduces the risk and costs associated with big jumps in technical change. Evolutionary product and process and organizational innovation is less expensive, because it requires minor adaptation of marketing, operations strategy and infrastructures. Markets recognize and sometimes even trigger such slow changes. Many organizations almost continuously produce such small adaptations to environmental changes of their output and/or their structure. Large changes, both in biology and economics, may produce monsters, which the environment often rejects as such, and they do not survive. The organization produces variety (at the level of technology, product, process, strategy and structure) in a bounded rational way, and the environment selects. Such slow process drives the organization and its technologies to local optima. 'Artifacts, like plant and animal life forms, can be arranged in continuous, chronological sequences. [. . .] Butler, Pitt-Rivers, Gilfillan, Ogburn and Usher all stressed the accumulation over time of small variations that finally yielded novel artifacts' (Basalla, 1988: 24). Yet the author recognizes that short periods of rapid change may exist between long periods of slow change and stasis. However, the vast majority of authors on technology have adhered to an evolutionary perspective.

In economics, Nelson and Winter (1982a) have identified one of the sources of slow change: the firm's routines, which are the genes of organizations. Over time, organizations have developed ways of solving their production and marketing problems; such a learning process has been long and costly, and has been reinforced by the building of complementary infrastructures and practices. Maureen McKelvey (1996) has presented the basic principles of evolutionary innovation in biotechnology. They include variation (generation of novelty); selection; transmission and retention of certain traits over time; and non-optimization but adaptation to local environments. Like Basalla (1988), McKelvey argues that biological evolution cannot be deemed identical to economic evolution. Nelson (2006) has also adopted this perspective: biological evolution and human culture share a few major unifying themes, such as variation, selection and retention, but are split apart by major differences, including the speed of change and the goal-oriented action of humans in cultural evolution. Also, within cultural dynamics there are large differences between fields, such as linguistic and policy evolution.

A lively debate among evolutionary economists and management theorists refers to the amount of inertia that organizations carry. At one extreme one finds the organizational ecology perspective, with such authors as Michael Hannan, John Freeman and Glenn Carroll, for which organizational inertia is predominant, and firm-level adaptation is limited. Populations of firms change by the birth and death of organizations; those that survive usually have the right genes from the start. Organizational ecology is more Darwinian, while Nelson and Winter (1982a) are more Lamarckian. The more the evolutionary approaches put the emphasis on the importance of strategy, including Nelson and Winter, the farther they are away from the organizational ecology perspective. Whatever the case, it is clear that most companies live and die with their original routines, technologies and strategies. These are the traditional small and medium-sized firms that Bhidé (2000) has shown to be the vast majority of firms (garages, hair salons, groceries, and so on). A few of them, usually medium-sized and large firms, tend from time to time to change their range of technologies, strategies and structures. This chapter adopts a mixed perspective: studies on firm mortality in all OECD countries show that the vast majority of firms (SMEs) disappear a few years after they were founded. A few of them manage to change, grow and adapt to their environment. Even among those that grow, adapt and change, many sometimes err in their choice of new routines, technologies and markets, and also disappear. The roads of industrial change of the latest years are littered with the remains of companies such as Northern Telecom and Nokia.

In this world of evolutionary innovation, technological trajectories abound, and technological discontinuities are amenable to modelling (Dosi, 1982). New technological paradigms (discontinuities) are linked to the emergence of Schumpeterian companies and the process of innovation stabilizes. The process is fairly structured: 'a technological paradigm (or research programme) embodies strong prescriptions on the directions of technical change to pursue and those to neglect' (Dosi, 1982: 152). Also, evolutionary innovation is the world of path dependency. Institutions, routines and technologies persist over time, even when they have outlived the social matrix in which they were born.

3.1.4 Innovation Cascades

The idea of innovation cascades is already present in Schumpeter:

> First, that innovations do not remain isolated events, and are not evenly dis-
> tributed in time, but that on the contrary they tend to cluster, to come about in
> bunches, simply because first some, and then most, firms follow in the wake of
> successful innovation; second, that innovations are not at any time distributed
> over the whole economic system at random, but tend to concentrate in certain
> sectors and their surroundings. (Schumpeter, 1939: 98)

More recently, a few authors have explored the subject without arriving at a complete explanation of the dynamics of its development. Delapierre and Mytelka (2003) link innovation cascades to the oligopolistic behaviour of large firms. Competition among large diversified corporations gener-ates the exploration of new technological domains, and the creation of new technologies and new industrial sectors. They do not make any link between their work and Schumpeter's, in spite of the obvious similarities. Antonelli (2008 and 2009) explains innovation cascades by the interplay of Marshall and Jacob externalities within clusters. Cascades appear in regional innovation systems, not necessarily in concentrated industries, as in Delapierre and Mytelka (2003). Lane (2009b) explains innovation cas-cades by a phenomenon called 'exaptive bootstrapping'. In biology, exap-tation is the use of a structure or feature for a function other than that for which it was developed originally through natural selection. 'Exaptation' is a change in the function of a trait during evolution. 'Bootstrapping' means to help oneself by one's own means and efforts. Thus, in the two previous explanations, the conscious efforts of economic agents launch a cascade; in Lane's approach, some agents would launch a cascade without even noticing it, just trying to solve a local specific problem. His example is Gutenberg's invention of printing using movable metal type around 1452–54. Such innovation launched a cascade where new organizational forms (printing companies), new technical novelties (new ink, paper), new markets (for printed books), and new functionalities emerge, and imitation from other economic agents increases both the market and the innovative activities, in a positive feedback dynamics that may extend over decades. Once it is launched, these self-reinforcing dynamics are difficult to control or predict, even for those that are actively involved in the process (Lane and Maxfield, 1996). Under such conditions, optimization and strategy making become difficult, if not impossible. And predicting technological trajectories is highly improbable. Finally, Berkers and Geels (2011) use the same notion of innovation cascades to describe a positive feedback innovation mechanism that has taken place among traditional small and medium-sized enterprises using innovations generated elsewhere (mostly equipment suppliers, but also government laboratories and universities). The authors make a passing remark about the fact that these cascades are different from those studies on scale-intensive and science-based industries and/or government utilities (Berkers and Geels, 2011: 243), but they do not cite any of the above-mentioned papers on innovation cascades. They contribute to the theory of technological transitions.

Technological transitions are major long-term technological changes. These technological transitions come along as a result of several mechanisms: niche-accumulation, technological add-on and hybridization (Geels, 2002). His idea of technological transitions is close to Schumpeter's approach of innovation cascades. Technological transitions occur in all different types of industries, from science-based to scale-intensive to government-supported sectors. However, 'transitions are characterised by one major, radical innovation or discontinuity' (Berkers and Geels, 2011: 230), while innovation cascades are more characterized by a stream of radical innovations.

In this chapter we contend, following Mokyr (2002), that innovation cascades in Western economies before the Industrial Revolution, such as the printing press, failed to promote sustained economic growth. They are different from present-day high-tech (information technology, nanotechnology and biotechnology) cascades. The reasons why innovation cascades before 1800 were short lived are many. First, the institutional environment did not contribute to their adoption but blocked the diffusion of innovation and the emergence of new radical innovations: indexes of prohibited books and censorship were widespread. Also, universities and private companies did not conduct R&D, and there were no public research laboratories to push the cascade further. Radical innovation depended on the individual efforts of remarkable luminaries like Galileo or Da Vinci. At that time, the innovation centres of the world were just a few cities such as Amsterdam, London, Paris and Venice, and within them there were few innovating organizations. Also, communications between those centres were slow and costly, and the scientific and technical knowledge of the times was scanty. Innovation came through serendipity, and was not the routine activity of many organizations as it is today.

After the Industrial Revolution innovation cascades became more frequent. One can find several of them associated with the rapid improvements in steel-making technology, the railway, the internal combustion engine, and chemicals, to name some of the most important industrial novelties in the 19th and early 20th centuries.

Post-war innovation cascades are increasingly frequent in Western countries. The reasons are many. For one, the stock of knowledge is growing by leaps and bounds. As a result, innovation, as measured by the number of patents and scientific publications, is increasing continuously. So the scientific and engineering raw material for innovation is today much more abundant (Kortum and Lerner, 1999; Larsen and Von Ins, 2010). Second, the rise of scientific collaboration (Greene, 2007) and particularly of international scientific collaboration increases the number of new combinations that may be produced on the basis of this new knowledge. The growth of international scientific collaboration may be explained by the diffusion of scientific capacity both within industrial countries and among emerging countries (Wagner and Leydesdorff, 2005). Also, rapid advances in communication and transportation technology increase the chances today that new combinations may emerge from international and inter-regional collaboration. Third, the institutional landscape has changed enormously: in each advanced industrial and emerging country, thousands of innovative firms and dozens of research universities, as well as public laboratories are now able to amplify and develop many technological trends in a way that was impossible 200 years ago. Thus, all these elements launch positive and self-reinforcing feedback processes that are increasingly unstoppable. Other key innovation institutions contribute today that did not exist in the 15th or 16th centuries, namely policy incentives, such as those aiming at the commercialization of university research, or pushing companies to conduct in-house R&D. These policies increase the likelihood that scientific novelty is used in industry and explored in universities and that it will launch an innovation cascade.

The previous world was one where technological trajectories and path dependencies were the name of the game. They still are numerous today, but innovation cascades, a world of self-reinforcing mechanisms, non-linear dynamics with many possible short-term equilibrium situations, make technological trajectories less evident than fifty years ago. Who could foresee the rise of the Internet, or the advances in computational genomics thirty years ago? Technological path dependencies also seem often to be interrupted by these innovation cascades. The dictum 'Natura non facit saltum' does not apply to these unpredictable cascades. The following table compares incremental and radical innovation, an important step towards defining innovation cascades (Table 3.2). It is important to recall that there is no universally accepted definition of either incremental or radical innovation.

Innovation cascades wreak havoc with the rigid sequence of PLC and ILC. Continuous new technological and industrial developments branch

Table 3.2 Incremental and radical innovation defined

Dimension of radicalness	Incremental	Radical	Authors
Impact on the industry	Low	High	Acemoglu and Cao (2015)
Source of subsequent innovation	No	Yes	Ahuja and Lampert (2001)
Older technology remains substitute for new	Yes	No	Arrow (1962)
Cost reductions	Low	High	Green et al. (1995)
Competitive advantage to adopters	Low	High	Kumar et al. (2000)
Benefits brought if successful	Low	High	Kumar et al. (2000)
Adoption risks	Low	High	Kumar et al. (2000)
Technical uncertainty levels	Low	High	O'Connor and Rice (2013)
Market uncertainty levels	Low	High	O'Connor and Rice (2013)
Resource uncertainty levels	Low	High	O'Connor and Rice (2013)
Organizational uncertainty levels	Low	High	O'Connor and Rice (2013)

out of the original products and services. In order to make the exact description of the series of radical innovation, the innovation cascade is redefined here as follows:

> An innovation cascade is a series of radical innovations that spans over a decade or more, and can be observed and measured through patents and scientific and technical publication.

And the hypotheses are drawn for our study as follows:

Hypothesis 1: An innovation cascade is emerging in the solar PV sector.
Hypothesis 2: Transnational technology diffusion has contributed to the innovation cascade in the solar PV sector.
Hypothesis 3: Different sources of demand promote the innovation cascade in the solar PV sector in different stages.
Hypothesis 4: The solar PV is a typical 'Mark I sector': it was born dispersed and its dispersion was maintained or even increased over time, a pattern that contradicts the PLC–ILC model.
Hypothesis 5: In science-based sectors, rapid changes in the science and

technology fundamentals modify the contours of the life cycle, in accordance with history-friendly models.

Hypothesis 6:   In science-based industries, such as the solar PV, demand changes modify the contours of the life cycle, in accordance with history-friendly models.

Our hypotheses are tested using the solar PV sector as the case. The chapter brings some aggregates about the rise of the solar PV sector, and then illustrates one of the major present-day innovation cascades with the growth of this high-tech sector. The growth of solar publication and patenting is also presented. A table with the different disciplines, application and key companies will help.

3.2   METHODOLOGY

3.2.1   For Innovation Cascade

We studied the solar sector with an emphasis on innovation and production. Among our most important sources of data, we used the United States Patent and Trademark Office (USPTO) patent database for the period 1976–2013. The solar sector is not an industry defined by NAICS or SIC codes, but a set of industries with different codes (Table 3.3). They include solar cells (the heart of the solar equipment), but also batteries, modules,

Table 3.3   NAICS and SIC codes of the solar photovoltaic sector

SIC code	NAICS code	Description
3674	334413	Manufacturers of Copper Indium Gallium diSelenide (CIGS) solar cells and solar foldable, flexible panels and off-grid glass modules*
3211	327211	Flat glass manufacturers
5074	423720	Plumbing and heating equipment and supplies
1711	238210	Plumbing, heating and electrical equipment contractors

Notes: * Products are crystalline silicon panels, thin-film panels, multi-junction panels, crystalline silicon cells, thin-film cells, multi-junction cells and organic cells and panels. In sum, their activities are the manufacturing of solar cells and solar panels. As of December 2014, the US industry had 5481 employees, 46 firms, and total revenues of US$1 billion. Annual growth for 2009–2014 was −7.3 per cent, due to international competition.

Source: US Department of Commerce.

inverters, advanced materials and increasingly specialized glass. The vast majority of the patents concerned solar cells, but we decided to search for USPTO patents having 'solar cell', 'solar glass' or 'solar battery' in the abstract. An initial manual search was revised by a computerized search and statistical analysis. Because the United States was the cradle of the sector and is still the most inventive country in the world, we take the American patent database as essential for our research.

The total number of patents for the different years has been calculated for the trends of the development of innovation. In addition, patents claimed from different countries and different regions have been studied.

In order to track publications, we used the Scopus scientific database with similar keywords as for patents. Up until 18 July 2015, there were a total of 111 173 such publications in Scopus. We used these databases to analyse the sector. The publications in the different countries and different regions were calculated.

In order to compare the publications and patents in different continents and show the effect of technology diffusion, we classify the countries with the publications and patents in solar PV in three categories: North America includes USA and Canada; Europe includes Germany, United Kingdom, France, Italy, Spain, the Netherlands, Switzerland, Sweden, Belgium, Russian Federation; Asia includes China, Japan, South Korea, India, Taiwan, Singapore and Malaysia.

We also analysed data from the US National Renewable Energy Laboratory (NREL) concerning the most important advances in solar cell efficiency from 1975 to 2014. Any organization claiming superior efficiencies in solar cells needs an external judge. The NREL, as the largest public research organization (PRO) in the world in solar technologies, is considered the major arbiter in this area, followed by the German Fraunhofer Institute, and the National Institute for Advanced Science and Technology (AIST) in Japan. The NREL list of best research-cell efficiencies was used as a way of distinguishing major inventions from the large list of solar cell patents.

As to production and markets, we used different sources such as the European Photovoltaic Sector Association reports, the Earth Policy Institute, NREL studies and specialized publications such as GreenTech Media and PVTech.

3.2.2 For Evolution of the Solar PV Sector

3.2.2.1 Definitions

In order to describe industry evolution as precisely as possible, the following concepts are defined:

- User innovators are the companies that directly benefit from the use of solar PV products, but also conduct R&D and produce patents from at least some of the results of their R&D investment in the solar technologies: companies such as Boeing, Canon, EXXON or Siemens are user innovators. Public laboratories and universities, as well as specialized small and medium-sized enterprises conducting research and patenting, or those that install panels for individual or industrial companies and do not conduct research are not user innovators. In our definition, users and manufacturers may coexist under the same roof and within the same enterprise (Block et al., 2016).
- Integrator: solar cells are the heart of solar photovoltaic systems, but are not the final product for most users. Other companies integrate the solar PV technologies into their products and then sell to the consumers; these include producers of solar roof panels, clocks, watches, pocket calculators, satellites, aircraft, solar tracking mechatronic equipment used in highways, telecommunications equipment and the like.
- Related diversification: the large companies usually spin off dedicated firms to produce solar PV products to end consumers in other applications.
- Mass market: the end consumers are individuals and commercial organizations that buy solar panels for houses, industrial firms or highways; also, companies producing portable electronic products such as clocks and pocket calculators.
- Niche market: the end consumers are in special industries such as aircraft, satellites or other specific products.
- Specialized manufacturer: these are manufacturers just focusing on solar PV cell or solar panels manufacturing, with no other focus or integration plan.
- Feed-in tariffs (FIT): 'A feed-in tariff (FIT) is an energy supply policy that promotes the rapid deployment of renewable energy resources. A FIT offers a guarantee of payments to renewable energy developers for the electricity they produce. Payments can be composed of electricity alone, or electricity bundled with renewable energy certificates. These payments are generally awarded as long-term contracts set over a period of 15–20 years' (US DoE, 2010). These FITs are used not only to create incentives for the adoption of solar PV technologies, but also for wind and other renewable energies. They were adopted in Germany, Italy, Spain, Japan, and lately in China. In the USA, six states have implemented such tariffs.

3.2.2.2 Sampling and data collection

In order to review the industry evolution of the solar PV industry in terms of innovation, the patent data in the USPTO are employed as the sample selection criteria. The data offered by the USPTO are selected because the United States assignees own nearly 50 per cent of the solar PV patents, and over 75 per cent of energy-storage patents. In addition, innovators based in other countries, mainly Japan, Germany, Taiwan or the People's Republic of China also patent their inventions in the United States in order to protect them from potential infringers.

The samples are grouped into three categories:

1. the earliest assignees in the solar PV;
2. the top 10 biggest patent assignee companies; and
3. the top 10 biggest manufacturers specializing only in solar PV cells and other solar system components such as solar high-tech glass and solar energy storage systems.

For each sample in the above groups we explore the information from their websites and from open information including the Internet, reports and journals to find the major business, the solar PV business and their marketing positioning.

3.3 RESULTS

3.3.1 Innovation Cascade in the Solar PV Sector

When the total number of patents and publications is studied, we observe that the growth rate of publications has begun to increase exponentially since 2000, and that of patents since 2010. Figures 2.1 and 2.2 show the exponential growth of innovation in the solar PV sector, one that we call an Innovation Cascade. Such growth started in the 1990s and keeps on rising today. This growth is partly due to government incentives for solar PV adoption, but also for solar R&D in private firms, academic and public research organizations.

When the technologies are studied, we observe several paths instead of one dominant direction from the cell efficiency evolution map from NREL (see Figure 2.3). Energy conversion efficiency is taken as the most important indicator to measure technology progress. According to Green (2009), over the last two decades, terrestrial cell measurements have evolved to the stage where independent laboratories measure the same result for standard silicon cells within a 1–2 per cent margin of error. As a result of early initiatives by SERI (Solar Energy Research Institute, now National Renewable Energy Laboratory, NREL), which encouraged the development of highly efficient silicon cells, several key silicon cell results were measured at NREL in the early 1980s, the beginning of what will be referred to as the 'modern phase' of silicon cell development. The calibration of NREL is often employed to show the progress of different categories of solar PV technologies. According to NREL statistics, the best research-cell efficiency evolved in the different paths.

There is neither a dominant standard nor a dominant design for the sector. When assessed based on cell efficiencies, the major evaluation measurement, radical innovations emerge from time to time. This looks very much like an innovation valley of the solar PV sector: the technological innovation fast-moving river evolves from rapids to waterfalls, and splits into several diverging flows that sometimes merge with other flows to form new estuaries. Thus, the hypothesis that there is an innovation cascade in the solar PV sector is supported. The hypothesis that the solar PV sector is characterized not by a decline of innovation but originally by a constant and even by a rising stream of innovations is strongly confirmed.

3.3.1.1 Innovation cascade due to the diversity of the innovation organizations

It is found that private companies initiated the most (nine) solar cell innovations (see Table 3.4). Also, PROs, SMEs and universities in different countries have initiated technological trajectories over the past half-century. Our patent analysis shows that different organizations contributed simultaneously to the innovation pools (see Table 3.5).

Judging by the total number of patents, large companies have been granted close to 60 per cent of solar cell patents, and small and medium-sized enterprises produced some 35 per cent of them. Universities also seem to play a major role in this high-tech fledgling sector (six), while some PROs (NREL) contribute significantly.

Large and small companies The contribution of large firms has been fairly constant over the years. In fact, large companies such as AT&T (Bell Labs), but also Boeing, DuPont, IBM, Kodak, Mobil Solar, RCA and Westinghouse were among the first to enter the solar race (Perlin, 2002). This pattern was even more marked in Japan with early large entrants such as Canon and Matsushita, and later ones such as Mitsubishi, Sanyo, Sharp and Sumitomo. In Germany, among the early entrants one finds large companies such as Robert Bosch, Siemens and Telefunken. According to pv-tech.org, today global leaders are large companies including Yingli Green Energy (China), Trina Solar (China), Sharp Solar

Table 3.4   Initial innovative organizations in solar cells (1975–2014)

Type of solar cells	Original organization	Country	First year cell tested
Amorphous ScH (Stabilized)	RCA	USA	1976
CGIS (thin film cells)	University of Maine	USA	1976
Single crystal, single junction cells (GaAs)	IBM	USA	1977
CdTe thin film cells	Matsushita	Japan	1977
Single crystal SI cells	Mobil Solar	USA	1977
Two junction cells	North Carolina State University	USA	1983
Single crystalline Si cells	Stanford University	USA	1984
Microcrystalline cells	Solarex	USA	1984
Dye sensitive cells	École Polytechnique Fédérale de Lausanne	Switzerland	1991
Three junction cells	NREL/Spectrolab	USA	1999
Organic cells	University of Linz	Austria	2001
Silicone heterostructures Crystalline Si cells	Sanyo	Japan	2001
Three junction cells (MM)	Spectrolab	USA	2003
Thin film crystal GaAs cells	Radboud University	Netherlands	2005
Organic tandem cells	University of Dresden	Germany	2008
Quantum dot cells	NREL	USA	2010
Perovskite cells	École Polytechnique Fédérale de Lausanne	Switzerland	2013

Source: Green (2015).

(Japan), Canadian Solar (Canada/China), Jinko Solar (China), Rene Sola (China), First Solar (USA), Hanwha SolarOne (China), Kyocera (Japan) and JA Solar (China).[2]

Table 3.6 shows that large firms (over 500 employees) dominate technology invention in the solar PV sector. This is true for all major metropolitan areas except Taiwan, where the three largest clusters have the Industrial Technology Research Institute (ITRI) as the main owner of solar PV technology. Since the 1980s, ITRI has become a major inventor in the field of semiconductors, and solar cells are specialized semiconductors. A capability that ITRI developed in one sector of the microelectronics industry could be transferred to another sector of the same industry, for a different application.

SMEs' performance in terms of their contribution to innovative outcomes varies by country. US SMEs represented nearly half of the US

Table 3.5   The major players in the sectoral system of innovation

Country	Universities		PROs		Private firms		Total	
	T. Pat	Key	T. Pat	Key	T. Pat	Key	T. Pat	Key
USA	78	14	60	31	2098	114	2236	159
Japan	11	0	23	1	957	22	991	23
Germany	8	4	22	7	353	5	383	16
Taiwan	31	0	59	0	130	0	220	0
S. Korea	11	0	39	2	160	2	210	4
France	0	0	6	0	32	0	38	0
Switzerland	2	5	0	1	34	0	36	6
Canada	0	2	0	0	35	0	35	2
China*	6	0	0	0	29	0	34	0
Netherlands	2	4	5	0	25	2	32	6
UK	0	0	0	0	30	0	30	0
Australia	15	17	0	0	7	0	22	17
Sweden	0	0	0	0	19	0	19	0
Austria	0	3	0	0	12	0	12	3
Italy	0	0	0	0	8	0	8	0
Total all countries	164	49	208	42	3929	145	4401	238

Notes: Australia's University of New South Wales does not patent under its own name.
Sums sometimes do not add up because of multiple assignees on the same patent.
* China includes Hong Kong.

Sources: USPTO and NREL.

companies that have been granted USPTO solar patents since 1976; they obtained about 14 per cent of them. In Japan, SMEs were non-existent among the innovators. In South Korea, with a similar industrial structure to that of Japan, chaebols hold the vast majority of US patents on solar cells. Smaller firms are particularly active in Germany, the United States and Taiwan. Few of them are key or prolific innovators in other countries.

Universities and the solar sector   Out of some 4000 US patents granted for 'solar cells' from 1976 up to 1 January 2014, only about 160 were granted to universities (4 per cent). But these (low) figures may be somewhat misleading. NREL has published a study on major increases in solar research cell efficiency over the last 40 years, and identified the organizations (academic, companies and PROs) responsible for such jumps. Out of 213 such events from 1975 to 2014, 49 (23 per cent) correspond to

Table 3.6 Solar cell PV patents by metropolitan areas and type of assignee

Metropolitan areas and prefectures	Large firms	Small firms	University	PROs	Individual inventors	Number of patents
San Francisco	219	124	15	10	1	369
Los Angeles	190	32	9	7		238
Greater Boston	77	27	21	1		126
Washington, DC	6			88		94
New York, NY	70	8				78
Princeton, NJ	50		5			55
Albuquerque, NM	52			2		54
Delaware Valley	38		14		1	53
Seattle	52			1		53
Dallas	43			2	3	48
Tokyo Pref.	216			1		217
Nara Pref.	209			1	1	211
Kanagawa Pref.	184		1	1	2	188
Kyoto Pref.	151			1		152
Osaka Pref.	105			2		107
Hyogo Pref.	73			2		75
Shiga Pref.	48					48
Munich	99	5	2	16		122
Frankfurt/Rhine	42					42
Seoul	128			15		143
Taipei	16		16	42		74
Hsin-Chu	9		11	28		48
Taoyuan/Zhongli	7		3	32		42
Total	2084	196	97	194	8	2637

Source: USPTO.

universities; the University of New South Wales (UNSW) was leading, with 17 efficiency records, thanks to its prolific School of Photovoltaic and Renewable Energy Engineering (SPREE) founded in 1977 under a different name. UNSW does not patent under its own title, but transfers technology to the university commercial arm, independent firms and spin-offs, which have patented the technology by themselves. BP Solar, the solar energy arm of the British oil company BP, and the Chinese solar company Suntech Power, founded by a graduate of UNSW, are among the transferees of SPREE technology. The École Polytechnique Fédérale de Lausanne

(EPFL) followed with five events. Holding three records each were Georgia Tech, the University of South Florida in the United States, Linz (Austria) and Radboud University in the Netherlands. In all, American universities had 11 such events, Swiss universities had five, German and Dutch universities four each. In addition, several universities produced spin-off firms, including the Laboratoire d'Énergie Solaire et Physique du Bâtiment at EPFL, Stanford University in the United States, the Technical University of Dresden in Germany and the University of New South Wales in Australia. Cases of technology transfer from university to sector were also frequent, with EPFL and UNSW in the lead.

Public research organizations (PROs) In the NREL study about key milestones in the efficiency progression of solar cells, PROs also occupy a prominent place, with 20 per cent of events. NREL, a US research organization based in Colorado and funded by the Department of Energy, with over 1600 employees and close to 700 visiting researchers, interns and contractors, with an annual budget of US $271 million in 2014, gets the largest number of events (30 out of 44 PRO events). NREL hosts a National Center for Photovoltaics, whose mission is to make solar energy competitive with any other energy source by 2020. NREL started in 1977 as the Solar Energy Research Institute. Since its inception 37 years ago, the cost of solar energy has declined by 96 per cent. NREL transfers technology to the sector and the general public through different channels, including licensing, contract research, spin-offs, publication and conferences. Its patents are held by the Alliance for Sustainable Energy, based in Golden (CO), close to NREL.

Indeed, NREL is the most advanced United States PRO in solar energy. German institutes follow: The Fraunhofer Institute for Solar Energy, with 1300 employees, is the largest European PRO in the area of solar energy. Located in Freiburg, Germany, it conducts research on materials, semiconductor technology, optics and photonics, chemical engineering and other related areas. It is responsible for four major events in solar energy, but none of them is at the origin of a new type of solar cell. Only NREL has been involved in a large number of such major events. A smaller institute, the Zentrum für Sonnenenergie und Wasserstoff-Forschung, established in Baden-Württemberg in 1988, has produced three events with its labs located in Stuttgart, Ulm and Widderstall. Yet none of these was the launching pad for a new technical trajectory.

In Taiwan, government research laboratories ITRI and INER have both built up a level of capability in PVs and this has been passed on to Taiwanese firms entering the sector, either in the form of transferred technology or (mainly) in the form of skilled and trained technical staff.

ITRI established a PV Technology Centre in 2006, but its involvement in the sector goes back at least two decades prior to this, to 1987, when its Energy and Mining Research Division first began R&D on both monocrystalline silicon and amorphous silicon. Indeed, Taiwan's very first company involved in SCs, Sinonar Amorphous Company, was set up in 1988 employing ITRI technology and founded by two former ITRI staff members. Today, ITRI holds nearly 20 per cent of PV patents owned by Taiwanese assignees.

3.3.1.2 Innovation cascade due to transnational technology diffusion

The US is still the scientific leader in solar cells, but Japan, South Korea and China are catching up. US leadership is also manifest in the number of patents. But Japan, Germany and Taiwan are not far behind; following in a third cohort are India, the United Kingdom, France, Australia, Italy, Spain, and several other countries. US innovation leadership is also evident through the study of rival technologies: Figure 3.2 shows that most competing technological trajectories of solar cells occur within the United States, the world's largest innovator. If the publications and the patents are studied overall, the US is the biggest innovator, China was the second largest in terms of publications and Japan in terms of patents, Germany, Taiwan and South Korea followed, and the other countries are distant competitors (see Figures 3.1 and 3.2).

The transfer of some scientific action to Asia is clearly shown by the data of Figures 3.3 and 3.4. In terms of publications, there were three peaks in the last 16 sixteen years:[3] in the 1990–97 period, inventions made in North America including USA and Canada dominated the sector; between 1997 and 2007, Europe became the leader; after 2007 until now, Asia has led the publication trends, which perfectly corresponds to the patent trends. For the US patents, innovation in Asia has kept on growing since 1983 and European patents are still at a low level. Today, North America and Asia host the largest numbers of patents owners. Transnational technology diffusion is confirmed, and both the original and catching-up countries are contributing significantly to the innovation cascade. Hypothesis 2 is supported.

3.3.1.3 Innovation cascade due to the different demands in the different stages

When the patent assignees in the different countries and regions are studied, we can find several periods in their evolution.

The first one was – after the creation of the solar cell converting sunlight into electricity in Bell Labs in 1954 – its application in satellites and aerospace products that require out-of-grid electrical supply. The United States

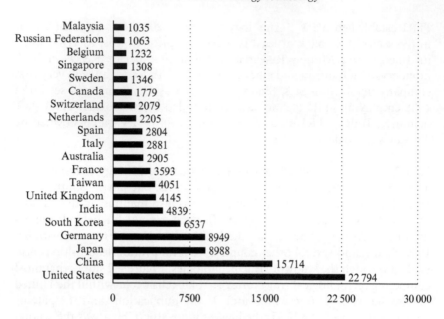

Source: Scopus.

Figure 3.1   Solar photovoltaic publication in main countries (1955–2015)

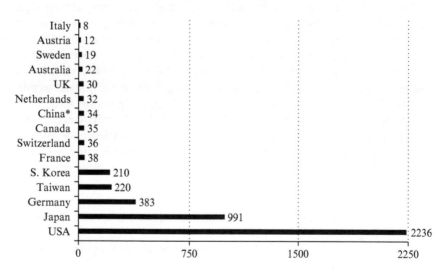

Source: USPTO.

Figure 3.2   Number of US patents by assignee country (1976–2013)

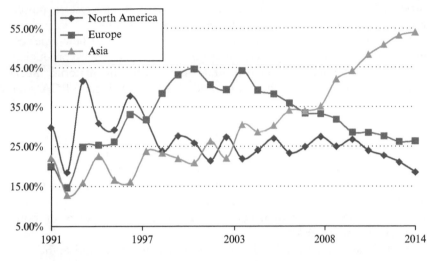

Source: Scopus.

Figure 3.3   The percentage of world publications on the different continents

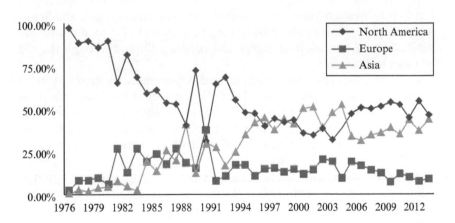

Source: USPTO.

Figure 3.4   The percentage of US patents on the different continents

hosted the main inventors and innovators in companies such as Hoffman Electronics, Signal Corporation, Communications Satellite Corporation and Raytheon. The first solar cell powered satellite was launched in 1958 and was a total success (Perlin, 2002). At this time, specialized companies produced most solar cells and innovation. The cost of such electric power was very high, as the efficiency of these solar cells was low.

In a second wave of innovation, in the 1960s and 1970s new types of innovators appeared; they were user innovators, companies producing solar cells for pocket calculators and other electronic material. Such innovators included Canon, Sharp and Toshiba in Japan, RCA and Texas Instruments in the United States. At the same time, some large oil companies, such as Exxon, and then ARCO in the United States became interested in solar energy for powering offshore oil and gas exploration and production. In Europe, Telefunken and other firms involved in the production of satellites became interested in solar PV technologies and conducted R&D and innovation. Australia also invested in solar PV in order to power telecommunications equipment in remote zones. The new wave of applications brought new innovators, most of them being both users and manufacturers.

As innovations accumulated, solar cell efficiency grew and additional innovators launched their products in the market, the cost of producing solar energy declined. A third wave of innovation arrived in the late 1980s and 1990s; in this period, early applications boomed. Rooftop solar panels and the first solar power plants connected to the grid appeared in California. Western European countries such as Germany, Italy and Spain adopted policy incentives, mostly feed-in tariffs (FIT), to increase the production of photovoltaic energy and reduce pollution through the use of clean technologies.

In the present wave, the fourth, several Asian countries such as China, Taiwan, Japan and South Korea have taken up the baton. The European countries have reduced their support for the adoption of solar technologies, and curtailed their innovative effort. Not only do East Asian countries manufacture the majority of solar PV equipment today, but they are also implementing incentives to produce more photovoltaic energy at home. However, the United States is still the most innovative country. China and Taiwan do not produce the most advanced solar equipment but rely on economies of scale to reduce cost; thus the rapid increase in adoption of inexpensive rooftop and grid-connected solar panels 'crowds out' the most advanced, but more costly, US-produced equipment. However, the US Department of Energy (through its National Renewable Energy Laboratory) predicts that in 2020 solar PV energy will be competitive almost everywhere with conventional sources of energy (US DoE, 2010). It is already competitive in most areas of Spain, Portugal, Italy, California, Texas and large portions of Africa, Mexico, Central and South America (see Table 3.7).

In addition to these user-manufacturer innovators, universities and

Table 3.7 Four periods in the development of solar PV technology

Period	Years	Most innovative nation	Most innovative organizations	Cases	Users
The birth of the technology	1950s–1960s (lunar landing programme)	United States	Dedicated solar cell producers	Hoffman Communications Satellite Corp.	Satellite producers in the USA, Canada and Europe
The first commercial applications	1960s–1970s (some demands that the conventional energy cannot be realized)	United States, Germany, Japan	Mainly user-manufacturers, but also public labs and universities	Canon, Seiko, Sharp, Toshiba (JP), ARC, Boeing, Exxon, RCA, Texas Instruments (USA), Telefunken (DE)	Pocket calculator producers; offshore oil and gas companies; Australia Telecom
Large-scale applications	1980–1990 (mass market is emerging)	United States, Germany, Japan	User innovators manufacturers, dedicated solar equipment manufacturers; public labs and universities	MiaSole, Solarex, Solopower SunPower Corp.	Rooftop users, first large grid applications, electronic equipment
Wide adoption of solar PV and entry of new component innovators	2000– (mass market is expanding)	United States, Germany, Japan, South Korea, China, Taiwan	User innovators manufacturers, dedicated solar-equipment manufacturers; public labs and universities	First Solar, Solo Power, Solarex, Evergreen Solar, Solaria (US), Samsung, LG (SK), Canon, Sharp, Toshiba (JP)	Large grid applications, space, electronic equipment, rooftop users

government laboratories have for over sixty years contributed to the development of solar PV technology. If in the 1970s and 1980s the University of Delaware was a major contributor, later on the National Renewable Energy Laboratory in the United States, the Fraunhofer Institute in Germany, the University of New South Wales School of Photovoltaic and Renewable Energy Engineering and its ARC Photovoltaics Centre of Excellence, the MIT Photovoltaics Research Laboratory and others have become more prominent.

The solar PV sector is undergoing phenomenal growth in several countries. In 2014, those with the highest number of renewable energy jobs included China, Japan, the United States, India and Germany. China was number one in terms of solar industries employment with 1.7 million people, followed by Japan, with 377 000. The United States had over 174 000 people employed in different solar industries, according to the US Solar Energy Industries Association. The number is increasing by 20 per cent a year, while the number of people employed in fossil fuels energy production would tend to decrease. Yet solar PV activity is moving towards Asia while declining in Europe. Yet Germany still has 100 000 employees in solar industries, followed by France with 60 000 and Italy with 45 000. In the world, there are 2.8 million people employed in solar industries.[4] The sector is the largest employer in renewable energies in the world.

Thus our third hypothesis has been supported.

3.3.2   Evolution of the Sector

3.3.2.1   More diversified evolution path
By studying the earliest and current patent assignees (Tables 3.8, 3.9, 3.10 and 3.11), it is found that there are different uses of solar PV technologies in an increasing number of products with the development of the sector such as:

- Satellites, aircraft, oil and gas offshore production facilities;
- Calculators, watches and other portable products;
- Flat roof rigid panels for individual houses or industrial plants;
- Sun-tracking solar systems using mechatronics;
- Grid-connected systems (solar parks, photovoltaic power stations);
- Green solar cities;
- Emerging technologies such as concentrator photovoltaics (CPV) using curved mirrors to increase efficiency;
- Floatovoltaics: floating panel systems, to save valuable land;
- Grid integration systems, allowing the use of solar PV energy when needed and available;

Table 3.8   The inventors of solar cell technologies up to 1976

Assignees up to 1976	The first patent in solar cell	Year of foundation	Specialized industry	Roles in the industry	The number of patents up to 1976
Communications Satellite	1973	1962	Satellite	User	12
Raytheon	1973	1922	Defence contractor	Integrator	9
US Government	1975			User	7
RCA Corporation	1976	1919	Electronics (ceased operation in 1986)	Integrator	5
Texas Instruments Incorporated	1976	1951	Electronics, semiconductors	Integrator	3
Dow Corning Corporation	1975	1943	Chemicals, silicon derived polymers	Integrator	3
Kabushiki Kaisha Toshiba	1974	1875	Electrical equipment	Integrator	2
Seiko Group	1974	1881	Instrument, watch	Integrator	2
Motorola, Inc.	1973	1928	Telecommunications	Integrator	2
Alcatel-Lucent	1976	1872	Telecommunications equipment	Integrator	2
Hughes Aircraft Company	1976	1932	Aerospace and defence	Integrator	2
Bell Lab	1976	1925	Telecommunication	Integrator	2
The Boeing Company	1976	1916	Aeroplanes, rockets and satellites	Integrator	1
General Electric	1976	1892	Power generation	Integrator	1
Lockheed Martin	1972	1912	Aerospace	Integrator	1
Rockwell International	1972	1928	Mechanic and aircraft and satellite	Integrator	1
Dresser Industries, Inc.	1976	1880	Energy and natural resources	Integrator	1
Westinghouse Electric	1976	1886	Nuclear power company	Integrator	1
Licentia Patent-Verwaltungs-	1975	N/A	N/A	Integrator	2
New England Institute, Inc.	1976	N/A	N/A		1
Ragen Semiconductor	1976	N/A	N/A		1
Sensor Technology, Inc.	1976	N/A	N/A		1
Beam Engineering, Inc.	1976	N/A	N/A		1

Table 3.9   Top 10 patents assignees and specialized mass-market manufacturers

Top 10 user firms in terms of patents in solar PV	Number of patents in solar PV until 2013	Top 10 solar PV firms in terms of patents in solar PV	Number of patents in solar PV until 2013
Canon Kabushiki Kaisha	207	SunPower Corporation	64
Sharp Kabushiki Kaisha	91	SoloPower, Inc.	46
Samsung Group	84	MiaSole	26
Applied Materials, Inc.	81	Solarex Corporation	21
E.I. du Pont de Nemours and Company	78	Evergreen Solar, Inc.	13
The Boeing Company	65	Mobil Solar Energy Corporation	12
Mitsubishi Group	64	Solexel, Inc.	12
Siemens AG	63	Solyndra, Inc.	9
Sanyo Electric Co., Ltd.	63	Solaria Corporation	8
Raytheon	52	Stion Corporation	6

Note: The US government is number 2 in terms of patent numbers. However, in this table we have removed the US government to leave just the company assignees and have included the patent assignee in eleventh place in the list.

Source: USPTO.

- High-tech glass, increasing the efficiency of the panels;
- Solar energy storage batteries and accumulators.

Also, a separate study that we conducted on US patents in energy storage between 1976 and 2015 found that this particular industry differs somewhat from the other industries in the solar PV sector. While the solar glass industry is dominated by the very large producers of flat advanced glass which are mainly in the US, and the solar cell industry is full of global competition consisting of at least fifty companies of different sizes and countries, the energy storage industry is dominated by very large firms of different industries and technology assets (car producers such as Tesla, Ford and GM, and multi-technology corporations like Bosch, LG, Samsung and Siemens), which means that the

Table 3.10 *The role of top 10 assignees in the solar PV industry*

Top 10 assignees	Roles in the sector	The industry as the user	Function as suppliers	Market specialized in the solar PV industry
Canon Kabushiki Kaisha	User innovation	Camera		
Sharp Kabushiki Kaisha	From user-innovation to related diversification	Radio		Mass-market
Samsung Group	Related diversification			Mass-market
Applied Materials, Inc.	Supplier		Semiconductor equipment provider	
E.I. du Pont de Nemours and Co.	Supplier		Higher performance materials	
The Boeing Company	Niche-market manufacturer	Satellite		Highest-efficiency multijunction solar cells for spacecraft power systems
Mitsubishi Group	User-innovation to related diversification	Satellite		Mass-market
Siemens AG	Related diversification			Mass-market
Sanyo Electric Co. Ltd.	Related diversification			Mass-market
Raytheon	User-innovation to mass-market suppliers	Apollo Guidance Computer	Power storage and control	

competitors in the energy storage industry are different from the other two industries.

The solar sector thus includes several different industries, and is characterized by a large degree of variation and increasingly complex products. If at the beginning, solar panels were covered with ordinary glass, and were fixed and unable to track sun movements, today they are becoming increasingly complex products. Most of these products are supported by the state through national research institutes, academic research, FIT, and the usual panoply of financial support of OECD and emerging country governments for R&D and innovation.

Table 3.11   Top 10 specialized manufacturing assignees of the solar PV patents

Top 10 specialized assignees	Founding year	Sources of the technologies
SunPower Corporation	1985	Star scientists
SoloPower, Inc.	2005	From related companies
MiaSole	2004 (but became a member of the Hanergy group in December 2012)	From related companies
Solarex Corporation	In 1973 (but acquired by Amoco in 1983)	From related companies
Evergreen Solar, Inc.	1994 (bankrupt in 2011)	Not available
Mobil Solar Energy Corporation	Disappeared	Not available
Solexel, Inc.	2005	Not clear (no directly related sources)
Solyndra, Inc.	2005 (but ceased operation in 2011)	Not available
Solaria Corporation	2003	Not clear (no directly related sources)
Stion Corporation	2006	From related companies

3.3.2.2   Critics of the PLC–ILC theories

We have examined different approaches that analyse industry evolution. The most cited, and decades-old one, is the PLC–ILC approach, proposed in the 1960s by Raymond Vernon and developed by Steve Klepper in the 1990s. Both approaches argue that products are born in the richest countries – most often the United States – where the first imitators also appear. These products then adjust themselves to market conditions, through innovation, until a dominant design emerges. At this moment, product innovation starts receding while process innovation increases. The new product is exported to less affluent countries where a second cohort of imitators appears. Economic concentration rises and large firms dominate the industry. The entire industry tends to be delocalized to emerging countries such as China, where costs are lower than in the original innovating country. In these PLC–ILC models, as soon as the dominant design is widely adopted, innovation declines.

The sectoral innovation system proposed by Malerba (2002) builds on the PLC–ILC perspective and adds a few important dimensions. It argues that the institutions, the markets and the technological conditions under which they are born shape sectors. Sector perspectives are more convenient than industry ones: most modern complex products and services are composed of different industries. They are seldom composed of just one industry, narrowly defined by SIC or NAICS codes. In addition, other authors (Saviotti, 1996; Niosi, 2000) have noted the phenomenon of rapid product variation, a phenomenon that does not disappear in science-based industries and sectors, as argued by the PLC–ILC approach; on the contrary, variation increases over time, and so does product innovation. Product variation often requires the contribution of products and services from other industries. High-tech solar panels require advanced glass, and/or mechatronics sunlight trackers, while using a particular type of semiconductor (solar cells).

In addition, product variation produces industrial growth by the expansion of industries that participate in the new product (Metcalfe et al., 2006). Such expansion translates into aggregate economic growth. In the same direction, Saviotti and Pyka (2004) argued that economic growth occurs most often by the creation of new sectors.

3.4 CONCLUSION AND POLICY RECOMMENDATION

3.4.1 The Influencing Factors Model

Based on the evolution of the solar PV sector, we propose the influencing factors model (see Figure 3.5).

3.4.2 Policy and Managerial Implications

Our data analysis confirms hypotheses 4, 5 and 6 postulated in this chapter. The number of innovators and producers keeps expanding over time, and no dominant design is in sight in solar cells, with so many technologies competing for a rapidly growing market. The industry is producing an ever-increasing number of solar panel types, and the number of different batteries for these renewable energy systems also keeps growing. Innovation is expanding and has been growing for the last fifty years; no end is in sight as of today. However, some aspects of the PLC–ILC approach are substantiated: the industry moved from the United States to Europe and now is moving to East Asian countries. But innovation is not receding – quite the contrary – due to the support by governments and competition in the three sets of countries.

The policy and managerial implications are substantial. If the industry

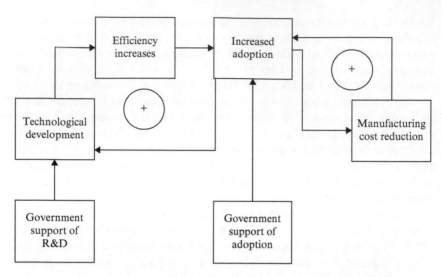

*Figure 3.5   The influence diagram of the solar PV sector innovation and
production system*

followed a rigid sequence, the policy opportunities would not be so
important. Conversely, if science-based industries followed a dense and
variegated bushy pattern instead of a linear arrangement, then policy
opportunities would be numerous. In the case of the solar sector, governments may decide to support one or another of the solar cell paths, or
participate in one of the components, such as solar cells, advanced glass,
new batteries, micro-grids, solar trackers or even software optimizing the
use of the solar equipment within large grids.

At the firm level, companies (and governments) have a choice: they
may try to position themselves as suppliers of key components for
product assemblers (that is, solar cells, solar glass, batteries), within
global supply chains, or enter in the adoption phases and become just
adopters.

NOTES

1. Sectors are sets of interrelated industries that usually collaborate to manufacture
    a product or a family of products (that is, automobile and tyre manufacturing
    industries).
2. Source: www.pv-tech.org.
3. Scopus data is not complete for the country of the authors before 1990 (for example, in
    1980, there is a total of 963 publications but there are just 177 publications identifying
    the countries of residence of the authors; in 1986, there are 913 publications, but there

are just 310 publications identifying the countries). To solve the problem, we just took the data after 1990 to note the publication trends.
4. UN Framework Convention on Climate Change, IRENA Report: *Renewable Energy and Jobs 2016*, Abu Dhabi, UAE.

4.   The catch-up of the Chinese solar PV sector

4.1   INTRODUCTION

Catching up is an essential part of the economic development process of countries behind the technological and economic frontiers, which involves learning and mastering ways of doing things in the leading countries of the era (Mazzoleni and Nelson, 2007). Practices in advanced economies do usually provide a model, but what catching-up countries achieve inevitably differs in various and important ways from the existing templates.

The debate about catching up was launched as a macroeconomic issue and was centred on productivity as the main indicator of catching up after the 1960s. Gerschenkron (1962) explicitly described the development problem of continental Europe during the second half of the nineteenth century as that of catching up with Great Britain. Abramovitz (1986) made the concept of catching up part of the standard vocabulary of development economists, and stimulated a number of empirical studies. Freeman (2002) used the examples of the development of the United Kingdom in the eighteenth century and the United States in the second half of the nineteenth century to underline the importance of continental, national and sub-national innovation systems for catching up. At the same time, the debate on catching up has moved from macroeconomics to the sectoral level. According to Kitschelt (1991), the description of sweeping aggregate national patterns may hide considerable policy variance across industrial sectors within each country, and the success of industrial strategies may depend more on sectoral governance structures than on national ones. Furthermore, national conditions constrain the learning processes of both industrial capabilities and governance structures. Sectoral and national-level conditions interact in shaping governance structures and innovation strategies, so besides national-level conditions, catching-up study on the sectoral level is useful. Most of the sectoral-level catching up is related to the technology-based view (Porter, 1990; OECD, 1992; Hobday, 1995; Kim, 1997a and 1980).

Lee and Lim (2001) and Lee and Kim (2004) have identified three different patterns of catch-up: a path-following catch-up, in which the latecomer firms follow the same path taken by the forerunners; a stage-skipping catch-up, where the latecomer firms follow the original path, but skip some stages, and thus save time and investment funds; and a path-creating catch-up, which means that the latecomer firms explore their own path of technological development.

For the path-following catch-up, there are abundant research results. In-depth case studies of countries catching up in the production and use of particular technologies have been made (see especially Ames and Rosenberg, 1963; Habakkuk, 1962; Von Tunzelmann, 1978; and many others), and some of the international trade and growth models are established (see Posner, 1961; Gomulka, 1971; Cornwall, 1977; Dosi and Soete, 1988). All the findings put the emphasis clearly back on the historical context and the institutional framework within which the process of imitation/technological catching up takes place. It includes the importance of 'developmental' constraints, primarily economic (such as the lack of financial or natural resources) or more political in nature, the role of immigration (Scoville, 1951) and other 'germ carriers', the crucial role of governments (Yakushiji, 1986) and the role of historical accidents. But according to Arthur (1988), the path-following catching up characterizes the increasing returns associated with industrialization and development which make the conditions of development so paradoxical: previous capital is needed to produce new capital, previous knowledge is needed to absorb new knowledge, skills must be available to acquire new skills, a certain level of development is required to create the infrastructure and the agglomeration economies that make development possible. It is within the logic of the dynamics of the system that the rich get richer and the gap remains and even widens for those left behind.

There are not many cases of stage-skipping catch-up and path-creating catch-up so far, but obviously, the catch-up of the Chinese solar PV sector is not in the track of path-following catch-up; it should be in the scope of either stage-skipping or path-creating catch-up. China has been the country with the highest production of solar cells for several years since 2007 and listed as the biggest added market in the world since 2010. Furthermore, when several big companies in Europe closed their manufacturing of solar cells, Chinese solar PV cell manufacturers acquired the related facilities worldwide. It is very interesting to explore the development of the Chinese solar PV sector to get the exact understanding of the special catch-up.

Zhang and Gallagher (2016) found that the main drivers for PV technology transfer from the global innovation system to China are world market formation policy, international mobilization of talent, the flexibility of manufacturing in China, and belated policy incentives from China's government. Zhang and White (2016) found that global entrepreneurship can contribute to the development of a local ecosystem, in addition to their passive and involuntary role as a source of spin-offs. The early entry start-ups developed very well in terms of production capabilities by overcoming the great 'liabilities of newness', building an effective organizational capability and establishing the legitimacy of the private Chinese solar PV firm as a viable organizational form, both domestically and abroad. Luo et al. (2014) found that returnees positively influence patenting activity and also promote neighbouring firm innovation in the Chinese photovoltaic sector. They also discovered that firms with returnees in leadership roles do more patenting. Fu (2015) argued that till now, private firms have been the major force in undertaking R&D and transforming scientific results into production technologies and ultimately commercializing them for the market. Zhang et al. (2014) added that the growth of both solar PV manufacturing capacity and deployment in China followed a very erratic path. The most important reasons are events which shape the wider policy priorities of China's government. Secondary factors include the government's poor management of the policy interaction between the domestic solar PV manufacturing industry and the deployment of solar PV across the country, as well as policy learning within government. Zhang et al. (2013) and Zhao et al. (2013) found that China's solar PV power, which is barely cost-competitive, has benefited less from the law and relevant policies. It was not until 2009 when the government rolled out measures to boost its domestic solar market for the purpose of weaning the country's solar PV industry off dependence on the overseas market that the solar PV power market in the country started to grow rapidly.

Nearly all the catching-up processes were accomplished before 2011, and the later developments are aimed at maintaining leadership in terms of market position. The development of the Chinese solar PV sector before 2011 is a new phenomenon deserving major attention.

4.2   LITERATURE REVIEW

Some studies have probed in detail the key processes involved in catching up (Hobday, 1995; Kim, 1997a; 1998; Kim and Nelson, 2000) and some factors have been discovered to explain how the catching up happened, such as governmental support (Perez and Soete, 1988; Lee, 2005; Mazzoleni and Nelson, 2007), a reasonable level of productive capacity (Perez and Soete, 1988; Lee, 2005; Liu, 2008), sufficient endowment of qualified human resources in the new technologies (Perez and Soete, 1988; Lee, 2005), location advantages (Perez and Soete, 1988) and intellectual property rights regimes (Mazzoleni and Nelson, 2007). Here, the literature from these standpoints will be employed to study the Chinese solar PV sector.

4.2.1 The New Techno-Economic Paradigm

The term of techno-economic paradigm was introduced by Perez (1983) and it shows that technology diffusion often has many impacts across the economy and eventually also modifies the socio-institutional structures. Such a meta-paradigm is the set of the most successful and profitable practices in terms of choice of inputs, methods and technologies and in terms of organizational structures, business models and strategies, and it can bring a valuable opportunity for catching up. Five technological revolutions in 200 years including the industrial revolution in England started in 1771, the age of the railway, coal and the steam engine started in 1829, the age of steel, electricity and heavy engineering started in 1875, the age of oil, the automobile, petrochemicals and mass production started in 1908 and the age of information technology started in 1971 are appropriate techno-economics examples (Perez, 2011). According to Perez and Soete (1988), each new techno-economic paradigm required, generated and diffused new types of knowledge, skills and experience and provided a favourable environment for easy entry into more and more products within these systems. Paradigm changes have historically allowed some countries to catch up and even to surpass the previous leaders. Lee and Lim (2001) believe that by taking advantage of new techno-economic paradigms, some countries make rapid progress and save time because they achieved some leapfrogging or skip some stages or even created their own path distinct from the forerunners. Lee (2005) also stated that the arrival of a new techno-economic paradigm could serve as a pull factor for leapfrogging.

According to Perez and Soete (1988), the life cycle of such a techno-economic paradigm is composed of a series of interrelated technology systems. There are four phases in the technology life-cycle model: introduction, early growth, late growth and maturity (Perez and Soete, 1988; Lee, 2005).

One of the important reasons for studying the new techno-economic paradigm is that it can temporarily open a window of opportunity for catching up if some requirements are satisfied (Perez and Soete, 1988; Lee, 2005). This solar photovoltaic sector with the new techno-paradigm provides valuable opportunities for developing countries to catch up. So what it would take for developing countries to catch up in the early phases of growth?

4.2.2   Government Support

The importance of governmental involvement in the catching up of developing countries is not new (Perez and Soete, 1988; Archibugi and Pietrobelli, 2003; Lee, 2005; Mazzoleni and Nelson, 2007). The development of the solar PV sector worldwide has provided a rich experience in this domain in Germany and China (Grau et al., 2012).

Teubal (1997) has classified the industrial policies into two categories: Vertical Technology Policies (VTP) and Horizontal Technology Policies (HTP). The former targets one specific sector and supports one specific technology. The latter is aimed at supporting various classes of socially desirable technological activities (SDTAs), such as firm-based R&D and innovation, technological infrastructure (both 'basic' and 'advanced'), and the transferring and adoption of new technologies. Both are deemed effective in promoting SDTAs across sectors and technologies, their importance deriving from being central components of government inducement of technology-based structural change in a wide variety of conditions (for newly industrialized countries, NICs), including situations with scant capacity to identify strategic sectors or technologies.

Are HTP or VTP better for the catch-up countries in the new techno-economic paradigm? There are two risks with leapfrogging: the risk involved in choosing the right technology or standards and the risk of creating initial markets (Lee, 2005). As governments can play the important role of facilitating the adoption of specific standards and thereby influencing the formation of markets at the right time, public support from governments is crucial, especially for a new techno-economic paradigm. But in the introduction phase of the new techno-economic paradigm, nobody, including government, knows what is the suitable technology standard for sectoral development and how to promote the formation of the market because there is often a lack of previous experience. In these circumstances, it is better to get direct government 'help' in a horizontal way. It comprises governmental subsidies of all sorts, preferential interest rates, R&D grants, tax reductions, protective barriers, and any other form of direct or indirect absorption of knowledge that would otherwise represent a cost to the firm (Perez and Soete, 1988).

So here is our first hypothesis:

Hypothesis 1:   For catching up in the introduction stage of the new techno-economic paradigm, government support with HTP is more adequate.

4.2.3 Entrepreneurs Benefit from Transnational Technology Diffusion

Nelson and Phelps (1966) put forward the hypothesis that the size of the gap between the technology frontier and the current level of productivity in a backward country is closely dependent on the quality of human capital in the former nation. Welch (1975), Bartel and Lichtenberg (1987), Benhabib and Spiegel (1994), Foster and Rosenzweig (1995) and Castles and Davidson (2000) documented the role of human capital in facilitating technology adoption. As to human capital in the new techno-economic paradigm, more attention should be paid to the entrepreneurs who are benefiting from the transnational technology diffusion. There are very substantial implications for economic growth and development involved in whether a nation's scientific infrastructure leads to the emergence of numerous entrepreneurs with a foreign education background and is conducive to their involvement in the commercialization of their discoveries.

According to Perez and Soete (1988), much of the knowledge required to enter a technology system in its early phase is public knowledge available at universities, although many of the skills required must be invented in practice. This implies that, given the availability of well-qualified university personnel, a window of opportunity opens for relatively autonomous entry into new products in a new technology system in its early phases. This is more important for international technology transfer, especially when the diffusion of major new technologies is hampered in some of those countries by the heavy investment outlays in the more established technologies, the lack of commitment of management and the skilled labour force to these technologies and even by the research geared towards improving them. In this situation, when entrepreneurs with advanced university training abroad in specialized areas come back to their less-developed home country, they may become the seeds to promote industrial development in the less-developed country, so that catching up can be accelerated.

So here is the second hypothesis:

Hypothesis 2: In the introduction stage of the new techno-economic paradigm, entrepreneurship with educational experience in the developed country is one of the essential components for catch-up.

4.2.4 Production Capability vs. Technology Capability

There is no doubt that technology has played an important role in catching up. Many studies show that more than 50 per cent of economic growth in advanced countries stems from technological innovation (Grossman and Helpman, 1991). But the technology capability can only be improved by integrating knowledge in the production process. Perez and Soete (1988) stated that a real catching-up process can only be achieved through acquiring the capacity for participating in the generation and improvement of technologies as opposed to the simple 'use' of them. This means that being able to enter either as early imitators or as innovators of new products or processes requires an integration capability consisting of technology capability and production capability.

In order to better explore the drivers of catching up, some papers have made the distinction between production capacity and technological capacity. According to Bell and Pavitt (1992), production capacity covers the knowledge and organizational routines required to run, repair, and incrementally improve existing equipment and products, while technological capabilities involve the skills, knowledge and organizational routines needed to manage and generate technical change. It increasingly happens that the kind of activities that foster the accumulation of the latter involve specialized R&D laboratories, design offices, production engineering departments and other organizations. By measuring the degree of catching up separately in terms of world market shares and in terms of technological capabilities, Lee and Lim (2001) try to explain the different records and prospects of Korean industries in the national catching up. They found that the differentiation between production and technological capability can be used to explain why some industries have achieved a remarkable catching up or leapfrogging and continue to have good prospects for the near future, whereas others are facing serious difficulties after a certain level of catching up.

But the technological capability and the production capability are interactive and inseparable. Industrial development is the process of building technological capabilities through learning and translating them into product and process innovations in the course of continuous technological change (Pack and Westphal, 1986). From a strategic perspective, the task of the latecomer is to devise ways of catching up by securing access to the knowledge and technology controlled by advanced firms in advanced countries. This requires them to understand the character and driving forces behind the industrial dynamics that govern the spread and diffusion of industrial processes and technologies around the world (Mathews, 2006).

Liu (2008) found that the most important capability for Chinese sector development is that Chinese companies can rapidly integrate market knowledge, technology opportunity and alliance capability. Lee (2005) stated that the important elements of catching up were to enter new markets segments quickly, to manufacture with high levels of engineering excellence, and to be first-to-market by means of the best integrative designs. Lee and Lim (2001) added that although technological capabilities are one of the most important elements, among the many determinants of market competition, such as manufacturing efficiency, marketing, logistics and so on, success in market competition can earn the firm the extra revenues much needed for R&D investment. So it seems that integrative production capabilities, consisting of both technological capabilities and production capabilities, are very important for catching up.

Here the third hypothesis is drawn:

Hypothesis 3:   In the introduction stage of the new techno-economic paradigm, comprehensive production competence integrated with technological and production capabilities are important for industrial development.

4.3   THE CHINESE SOLAR PV SECTOR

4.3.1   Government Support before 2011

It is widely accepted that the development of the solar energy sector is strongly dependent on governmental supporting policies such as market support programmes, which are acting as the main driving force for the development of the PV sector by serving customer needs with competitive costs. Since the 1990s, several countries, especially in Europe, have set up market support programmes to create the corresponding market (Table 4.1).

Table 4.1   Comparing the industrial supporting policies in different representative countries

Policy categories	Germany	Japan	US	China
Feed-in tariff	Available	N.A.	Available in California	Available but not workable
Net metering system	N.A.	Available	Available	N.A.
Investment subsidy	Available	Available	Available	Available
Tax reduction	Available	Available	Available	Available
Rooftop programme	Available	Available	Available	Available just for pilot programme

Source: Revised based on Liao and Xu (2012).

Chinese vertical sector policies for the solar PV sector before 2011 have been as follows:

- In 1996, the Chinese Brightness Programme was scheduled to run until 2010 with the aim of providing 100 watts of PV electricity to about 23 million poor people who had no electricity at that time.
- In 2006, the government began to invest money in several solar energy projects such as the Township Electrification Programme and the rooftop programme in Shanghai and Wuxi. Also, the Renewable Energy Law was taking effect at the beginning of 2006. By law, China planned to increase its renewable energy consumption to a full 10 per cent by 2010 (which has been accomplished) and required grid operators to accept electricity from registered renewable energy producers. A fund was set up to offer financial incentives to encourage the development of renewable energy projects and some very clear penalties for non-compliance were included (Ma, 2012).
- On 24 July 2011, the Chinese central government set up the feed-in tariff (FIT). But when compared with the law in other countries,[1] the Chinese support appeared too weak. Due to the lack of detailed regulations on the duration, the regional variation, the project application process and how to access the electricity grid, the Chinese FIT law had a quite limited effect in promoting the Chinese PV sector (Liao and Xu, 2012).

When comparing the market support programmes with those of developed countries including Germany, Italy, Japan and the United States, we cannot find any advantages in the support policies in China, which are neither more innovative nor more workable than other countries before 2011 (Table 4.1). Without the establishment of a solid domestic market, 90 per cent of Chinese solar PV products were exported from 2005 to 2010, and then this figure declined after 2011 (see Figure 4.1).

But even without strong vertical sector policies (not to mention vertical technology policies), how can the Chinese solar PV sector start, survive and then thrive?

Fortunately, although there are no effective VTP at the national level, some regional innovation policies, which aim at supporting the development of high-tech industries in general, helped the development of the Chinese PV sector.

Until 2012, about 70 per cent of Chinese PV cells were manufactured in the province of Jiangsu. It is interesting to see what they have done in order to promote the development of the sector. Wuxi, where Suntech (the

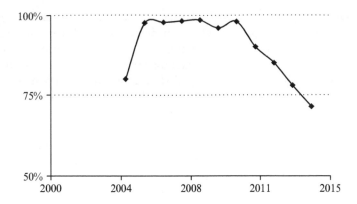

Source: Data before 2011 are from the China Chamber of Commerce for Import and Export of Machinery and Electronic Products (Yin, 2011); data 2012–2014 are from the Chinese Academy of Engineering quoted from National Energy Administration.

Figure 4.1 *The percentage of exports in Chinese solar PV equipment production*

biggest producer of the Chinese solar PV sector before 2012) is located, is in the province of Jiangsu. Wuxi, one of several industrial cities in the province, has its distinctive regional innovation system and horizontal innovation policies. Similar to the Shanghai Zhangjiang model,[2] the Wuxi innovation-promotion model is characterized by its openness in terms of the establishment of a friendly and warm microenvironment to encourage innovation and entrepreneurship: the performance is assessed not just based on the highest rate of GDP growth; it tries to establish links with the Top 500 entrepreneurs in the world and provide the Chinese abroad with capital to attract them to return to China; it creates seed funds and incubator funds with the intention of long-term development but not short-term profit; it encourages risk-taking and tolerates the failure of innovation; it provides a package of professional services including financing, law, accounting, headhunting, marketing, equipment leasing and retail; and it establishes the infrastructure for a high quality of life for high-tech talents and entrepreneurs. With the HTP packages, the Wuxi government has become a strong supporter of solar PV firms in the early stage, such as Suntech.

It is said that the establishment of Suntech has shortened the technology gaps between the Chinese solar PV sector and those in the developed countries for at least 15 years.[3] In the process of initiating Suntech, support from Wuxi local government is essential:

- Dr Zhengrong Shi received financial and social network support from the Wuxi government when he came back from Australia as a graduate student. At that time, he had just personally amassed US$400 000 as the initial capital for launching the business. But this amount is far from what is needed to launch a solar PV manufacturing factory; it is not even enough to buy the production equipment. In these circumstances, government officers not only persuaded the six local companies, Jiangsu Xiaotian Co. Ltd, Wuxi Guolian Trust Co. Ltd, Wuxi Shuixing Co. Ltd, Wuxi High-tech Investment Co. Ltd, Wuxi Venture Capital Co. Ltd and Wuxi Shanhe Co. Ltd, to invest US$6 million in cash in Suntech by taking a 75 per cent share of Suntech, but also persuaded all six investors to reward Dr Shi with US$1 600 000 for his patents and technologies as the corresponding shares in Suntech. All this help was so vital to Dr Shi that without the government's intervention, he would not have been able to start Suntech.
- The most valuable incentive was that, just before Suntech was listed on the New York Stock Exchange, the Wuxi government withdrew from the board of directors, giving more decision power to Dr Shi to play in the world arena.

The Wuxi government has provided other help including tax returns, financing, and good benefit packages for the recruitment of experts. What the Wuxi government has done is not uniquely related to the PV sector but is a common practice with Horizontal Technology Policies. In the initial stage of catch-up with technology choice and market formation risks, it is wise to provide the HTP in a wider scope to support all possible industries, and let the strongest win while some are eliminated by the market.

When the Chinese central government did not issue very strong supporting policies, the horizontal technology policies at the regional level provided most valuable support.

4.3.2 Human Resources

Crystalline technology (c-Si) (the first technological generation for the PV sector) accounts for more than 90 per cent of the actual PV systems in the market. The reason why it is so dominant is that it has used the technological and R&D efforts of the semiconductor industry for the electronics sector since the 1960s. All the Chinese PV companies adopted c-Si technology for cell manufacturing before 2011. In the process of technology transfer, entrepreneurs with education experience in developed countries who had been studying and working in developed countries played a very important role.

It is widely accepted that Suntech was the original technology base in the early years of the development of the solar PV sector in the 2000s, and most of the technology experts and engineers in the late-established companies have worked for Suntech.[4] Dr Zhengrong Shi, the former president and CEO of Suntech, is the key person not only as the founder but also as the technology developer. Actually, Dr Shi is a typical Chinese scholar-entrepreneur with experience of studying and working abroad. After obtaining his bachelor's and master's degrees in China, Dr Shi was sent by the Chinese government in 1988 to study at the University of New South Wales, Australia. He obtained a PhD in 1991 on innovation of polysilicon thin-film solar cell technology. As one of the best PhD students of Professor Martin Green, winner of the 'The Right Livelihood Award' in 2002, Dr Shi achieved performance excellence first in his studies and then in his work on thin-film technology innovation in Australia. During his career as executive director at the research centre in the university and Australia Pacific Power Co. Ltd, Dr Shi personally held six USPTO patents.[5] In 2001, Dr Shi returned to China and set up Suntech Power Co. Ltd. He knew the PV technology and the production of the modules so well that he had the confidence to buy second-hand production equipment from the US. He established working teams and guided the workers to manufacture the solar PV cells. At that time, all the technologies and the production capabilities of Suntech were technologically and efficiently superior to the other PV companies and this brought about a big change for the whole Chinese PV sector and also was the key factor for the later success of Suntech itself.

Dr Jianhua Zhao, classmate and colleague of Dr Shi's in Australia, with similar study and work experience to Dr Shi, also later founded and developed CSUN Corporation, a NASDAQ-listed leading manufacturer of solar cells and modules.

To some extent, it is that group of returning scholar-entrepreneurs who explored the opportunities and developed the sector in China. Their knowledge and experience obtained in developed countries was the most important factor for the catch-up of the solar PV sector in China.

It is estimated that there was a total of 400 000–500 000 workers in the PV cell and module manufacturing sector in 2012 in China, most of them from three sources:[6]

- Domestic researchers in universities transferring from relevant areas. Many universities have established research institutes on PV, for example, the Green Building and Energy Centre in Tongji

University, the Solar Energy Materials Laboratory in Guangzhou Institute of Energy at the Chinese Academy of Sciences, the Solar Energy Research Institute of Shanghai Jiao Tong University, the Solar Systems Research Institute in Zhongshan University, and so on. Most scientists in these research organizations transferred from related domains such as physics and materials science, which is becoming an important source of technological innovation of the sector.

● Employees transferred from the semiconductor and other related industries. In these related industries, the employees are well trained and have relevant knowledge, and experienced a similar mode of sector development and change. Not only at the beginning but also in the later stages, are more and more skilled workers in the semiconductor, electronics and other industries moving to the photovoltaic sector since the sector has become more and more attractive.

● Technicians and workers from school–enterprise cooperative training programmes. Many PV companies sought to begin cooperating with institutes of technology to acquire qualified workers; thus some colleges were set up: Suntech College in Wuxi Institute of Technology sponsored by Suntech and CSI Photovoltaic Technology College in Changshu Institute of Technology sponsored by Canadian Solar Inc. (CSI). There are some companies cooperating with local education institutions, although not in title, but with either a tacit consensus or signed alliances to get the students qualified for PV industries. Xinyu College hired some engineers and technicians from LDK Solar Hi-Tech Co., and Jiangxi Sun Optoelectronics Technology Co. has sent some employees as part-time teachers to train the students.

The review of the sector shows that it is the entrepreneur with education experience in developed countries that played a key role in initiating and developing the Chinese solar PV sector. With its spirit of entrepreneurship and risk-taking, the sector can develop and thrive, the demand for human capital can be created, and human capital can be accumulated in China, all of which are basic elements for catch-up.

4.3.3   Integrative Production Competence

The technological capability of Chinese solar PV is not strong. By establishing a new database of 19 105 solar photovoltaic patents in 1st, 2nd and 3rd generations of Taiwan, Korea and China at the USPTO over 24 years (1984–2008), Wu and Matthews (2012) analysed the knowledge flows revealed in these patents using a set of 12 International Patent

Classification technology categories, and they found that China still exhibits a low degree of patenting in the emerging new generations of technology.

The great pressure involved in manufacturing with low costs has impeded the pace of improving solar energy transforming efficiency, which is the key measurement for technology progress in the solar PV sector. Even though the first generation of crystalline technology is far inferior to thin-film technology in terms of transformation rate and the requirement of silicon as raw material, even though entrepreneurs with education experience in developed countries like Dr Shi have expertise and even patents in thin-film (the second generation of solar PV technology) themselves, they cannot make their products by using technologies in the second generation. Actually, when Suntech was already big enough to try the second generation, all the efforts of Dr Shi, who had the technological expertise and professional preference to develop the products in the most updated technology in thin-film technology, failed.

- After he established the production line on thin-film technology in May 2007 in Shanghai, the financial crisis in 2008 reduced the price of polycrystalline as the raw material of crystalline technology from US$400 to US$50. This greatly reduced the manufacturing cost of the first generation technology, which made manufacturing with this technology yield much greater profits than the second generation. At the same time, Applied Materials Co. in the US, whose products are regarded as the most suitable equipment for manufacturing thin-film products, declared that they were withdrawing from the business of thin-film solar cell manufacturing equipment in July 2010. For these two reasons, the efforts of Suntech in the thin-film sector did not bring any advantages; Dr Shi decided to change from thin-film technology to crystalline technology to recuperate the sunk costs.
- Suntech set up Sichuan Suntech with Sichuan University to deploy R&D on thin-film solar cells in 2009. But despite the investment of several hundred millions in RMB, there was no clear progress until 2012.

Adopting the industrial technologies is paradoxical. On one hand, a greater and greater supply of silicon has released a great deal of pressure from the raw materials for manufacturing crystalline solar cells and modules, which reduces the production cost of first-generation technology, making the cell price more competitive. On the other hand, the higher transformation rate has made the sector evolve into thin-film, that is, the second generation. The Chinese solar manufacturers are facing difficulties in upgrading the technologies because they lack a way to balance production cost and technology advancement.

Technology development is becoming more and more systematic, and the efforts of one or several Chinese companies cannot support and push the technology to evolve much further. But successful entrepreneurs with education experience in developed countries, especially those with a scientific background in advanced technologies, cannot contribute their technological advantages fully.

With their weak technical capabilities, Chinese companies tend to focus more on the technologies required to improve the efficiency of the manufacturing process than on the PV technology development itself. Innovations in the Chinese PV sector are mainly related to PV manufacturing equipment such as module laminators, wafer etch/bath and mono-crystalline wafer pullers, which aimed at technologically reaching the most efficient per dollar manufacturing process of the market, and neglected some important areas such as system components like inverters, batteries and control electronics which typically make up about 25 per cent of system costs.

It is reported that Chinese companies have had a history of being able to enter the market of different industries very rapidly and aggressively (Liu, 2008). They attained entry by technologically reaching the most efficient per dollar manufacturing process of the market, as well as combining a highly skilled cheap labour force and local low-priced manufactured automation equipment. By using turnkey equipment, cheap qualified labour and low administrative costs, Chinese companies were able to build large PV manufacturing plants and China became the biggest manufacturing country in the world within just six years of starting operations in 2001. The Chinese PV sector has seized the PV market opportunity by forming its productive capability quickly. This strategy has given China a great competitive advantage over Western companies in the new techno-economic paradigm.

4.4   CONCLUSION AND DISCUSSION

Based on the study of the Chinese solar PV sector up until 2011, the results of testing the three hypotheses have been examined:

- The first hypothesis states that in the introduction stage of the new techno-economic paradigm, government support with HTP is more feasible. This hypothesis was supported.
- The second hypothesis states that entrepreneurs with education experience in developed countries with an education background

abroad constitutes one of the most important components for catch-up, and it is the case in China.
- The third hypothesis states that integrative production competence together with technological and production capabilities are critical for industrial development. The third hypothesis is unproven. In the catch-up stage, only production capabilities are more important.

In fact, the situation of the Chinese PV sector has changed a great deal since 2011. With the European debt crisis worsening, the European photovoltaic market, which always depended on governmental support, has shrunk due to fewer governmental subsidies. The 70 per cent decrease in demand from Europe has brought even fiercer competition among Chinese solar PV manufacturers with lower prices and lower profits. On 8 November 2011, the US Department of Commerce officially initiated anti-dumping and countervailing duty investigations into Chinese solar cell exports to the US. SolarWorld, the US solar equipment manufacturer, asked the US government to charge Chinese exporters a 49.88 to 249.96 per cent levy in anti-dumping tax and countervailing duties, which makes the situation for China's photovoltaic sector even worse. With several unfavourable factors, the Chinese solar PV sector has entered a difficult stage.

According to Lee and Lim (2001), a sustained long-term increase in market shares is very difficult if it is not accompanied by increases in technological capabilities. If these firms try to increase their technological capabilities, they will find it more and more difficult and expensive to buy the more advanced technologies needed for higher-level market shares. Perez and Soete (1988) stated that the big problem is whether the endogenous generation of knowledge and skills is sufficient to remain in business as the system evolves, which requires not only constant technological effort but also a growing flow of investment. Development is not about individual product successes but about the capacity to establish interrelated technology systems in evolution, which generate synergies for self-sustained growth process.

Although Chinese PV companies have done well in the global markets, it is hard to see if the advantage in the initial stage can be maintained in the later phases. They may lose their market position very rapidly in the future as a result of a lack of necessary R&D input on thin-film or even the other future generations of the disruptive technologies that have a steeper learning curve and may change the market situation very quickly. So it can be concluded that catching up in production only is not enough; in fact, if the technological basis of the sector changes, the catching-up effect disappears. Even in a specific sector, the production leadership may be very unstable and will be lost without a sound effort on the innovation front.

But the Chinese solar PV sector has great potential to develop further because of the following:

- Considering that more than 70 per cent of Chinese solar PV products are exported to other countries, Chinese solar PV has a very big untapped domestic market. It is expected that increased PV presence in China would result in more experience for the sector itself, and efficiency and quality improvements would naturally follow. From 2011, some policies to stimulate domestic consumption are beginning to work well.
- Technology capabilities are on the way to being strengthened. According to Wu and Mathews (2012), in the new generations of solar PV (2G and 3G) technologies, China not only emerges as a leader in terms of local citation amongst the three main countries, but also tends to be more science-based, with 83 per cent of the patents owned by academia, all of which indicates that China could well be pursuing a leapfrog strategy straight to the newer technologies.

Whether China can maintain its market leadership and at the same time upgrade its manufacturing technologies is still in doubt.

NOTES

1. Germany introduced the law in 2004, which proved to be very effective and was regarded as the main driver for further cost reductions and a transition to economies of scale. Japan, the United States, Spain and Italy also issued regulations following Germany to support the photovoltaic industry, including a subsidy tariff provision and fixed the share percentage for electricity enterprises to buy out.
2. In China, three categories of regional innovation systems models are regarded as being effective: the Beijing Zhongguancun model dominated by the government and in which most R&D activities are conducted by governmental research organizations; the Shanghai Zhangjiang model promoted by the government and marketed together, in which companies actively conduct R&D integrated with the strong industrial base and the effective interaction in the Yangtze River Delta economy zone; and the Shenzhen model in the province of Guangdong, characterized by technology transfer from outside the company and then absorbed internally driven by the strong private sector innovative and entrepreneurial spirit.
3. http://www.china-apt.cn/news/news_show.aspx?id=1712.
4. A CEO's comments on the bankruptcy of Suntech, 22 April 2013, http://news.imeigu.com/a/1366636502643.html.
5. The patent numbers in USPTO are: US Patent 5942050, US Patent 6624009 B1, US Patent 6420647 B1, US Patent 6538195 B1, US Patent 6551903 B1, US Patent Application Publication US 2009/0007962 A1.
6. Workforce of the Chinese PV industry, 16 August 2011, http://www.360doc.com/content/11/0816/20/7197533_140887040.shtml.

5. Anchored clusters: the rise and fall of solar PV agglomerations

5.1 A THEORETICAL INTRODUCTION

High-tech industries tend to cluster in geographical regions. Different explanations for such agglomerations have been proposed, including the role of big corporations acting as magnets, such as large systems integrators (Perroux, 1972), externalities provided by many firms and institutions in innovative clusters (Porter, 2000, close to Marshall), knowledge-producing anchor tenants or large research-intensive corporations according to Agrawal and Cockburn (2003), or public research institutions in Feldman (2003). The national innovation system (NIS) approach has added its own perspective, calling these agglomerations 'regional innovation systems' (RIS) (Boschma, 2005; Cooke, 2001; Niosi, 2005). In a rare demonstration of unanimity, studies argue that high-tech firms in biotechnology, information and communication technology and nanotechnology tend to agglomerate in a few regions in each industrial or emerging country (Swann et al., 1998; Audretsch, 2001; Niosi and Bas, 2001; OECD, 2001; Niosi, 2005; Youtie and Shapira, 2008; Mangematin and Errabi, 2012; Kupriyanov et al., 2014). Anchor tenant firms and research institutions were most often considered the originators of these high-tech innovative clusters. The main reason for the agglomeration and the long-term evolution of these anchored clusters were basically knowledge externalities produced by these R&D-intensive organizations.

High-technology anchor tenants are identified as large R&D-intensive firms, or research universities and public research organizations (PROs), as defined by their patenting activity, with a strong focus on a particular technological field (Agrawal and Cockburn, 2003; Feldman, 2003; Link et al., 2003; Niosi and Zhegu, 2010; Schultz, 2011). Anchor tenants support patenting activities of both anchor and non-anchor firms in the cluster. Anchored clusters have been identified in information technologies, biotechnology, aerospace and nanotechnology. Box 5.1 includes some of the most accepted definitions of anchor tenants in the literature.

Critics have underlined the fact that these clusters are often poorly defined in both geographical and industrial terms (Amin and Robbins,

---

BOX 5.1   DEFINITION OF ANCHOR TENANTS

Korhonen and Snakin (2001): 'An anchor tenant is an influential organization in the system that drives its main material and energy flows. And hence can serve as the key actor in the environmental management effort of the system.'

Korhonen (2001): 'In this paper, a regional industrial ecosystem that relies on a power plant as its key organization, as an anchor tenant, is considered in the context of energy production and consumption.'

Link et al. (2003: 1218): 'An anchor tenant is a firm that generates positive demand externalities by attracting additional tenants and stimulating traffic within a commercial operation (typically a shopping mall or industrial park).'

Feldman (2003: 320): 'The Anchor Firms are more established firms with product lines that predate the biotechnology revolution but have current efforts involving biotechnology. More generally, regional anchors may encompass other institutions such as universities, government labs, research institutes and other entities.'

Niosi and Zhegu (2010: 263): 'The anchor tenant is an organization, often a large innovative firm or a research university or public laboratory that produces knowledge externalities in the region where it is located.'

Schultz (2011: 560): 'An anchor tenant is a firm traditionally heavily engaged in R&D with research interests in a technology being developed in the geographic area.'

1990; Martin and Sunley, 2003). What exactly are clusters: metropolitan areas, cities, provinces, states, or all of the above? How many different and/ or related industries does an agglomeration have to host in order to qualify as a cluster? Does a biotech cluster need only biotechnology and venture capital firms, research universities and venture capital? Is there a minimum threshold in terms of employees, sales or number of organizations necessary for being called a cluster? To what extent do these collocated industries need to be related in order to represent a cluster? Porter (2003) launched the idea and proposed a method of measuring relatedness based on employment. However, it was Boschma et al. (2012), based on Frenken et al. (2007), who used SIC and harmonized industrial system codes to measure relatedness. They found that related variety increased the chances of regional economic growth and resilience.

In addition, clusters often grow, but sometimes decline, and even disappear, without attracting much attention. Yet some scholars have analysed the rise of new clusters in competition with established ones (Maskell and

Malmberg, 2007; Shapira and Youtie, 2008; Menzel and Fornahl, 2009), as well as the decline of some of them (Hassink, 2010; Suire and Vicente, 2009; Østergaard and Park, 2015).

In anchor tenant regions, based on a large corporation, the anchor has a major influence on the dynamics of the cluster. It may, for instance, create new firms through spin-off formation and these spin-off firms perform better than other firms attracted to the cluster (Klepper and Thompson, 2007). Also, having a variety of innovative firms and knowledge-producing institutions seems a major factor for growth and resilience (Agrawal et al., 2014). However, the exit of the anchor tenant may play havoc with the cluster (Østergaard and Park, 2015). Conversely, in bottom-up RIS, established on a large number of small and medium enterprises, the withdrawal of venture capital support for new technology-based firms may also weaken the cluster. Rapid technological change and lock-in in inferior technologies can be a decisive factor in the decline of the regional innovation system, whether anchored or not (Boschma, 2005). In addition product life cycle (PLC) and sector life cycle (ILC) theories emphasize the fact that, as the product and the sector matures, a shake-up occurs that reduces the number of active firms, and tends to move the sector towards emerging countries, thus contributing to the decay of existing industrial regions in advanced countries. Such a process has been observed several times in the decline of the US rustbelt; Detroit, once the world capital of the auto sector, is a case in point. Audretsch and Feldman (1996) suggested that in the early stages of a sector, the propensity to agglomerate geographically is strong, but this propensity declines as the sector matures. This process may be unfolding in high-tech sectors such as ICT. Other authors have pointed to regional policy failures as important factors of decay: inappropriate policy designs may also explain the weakening of some regional innovation systems.

In summary, studies of high-tech RIS have recently focused not only on the birth and growth of these systems, but also on the factors that may weaken and eventually destroy them. These factors may be internal to the agglomeration, such as lock-in to inferior technologies, the withdrawal of venture capital or the exit of the anchor tenant, but also external such as the attraction of the anchor to other regions, or the migration of technology to new regions in a PLC–ILC process. Similarly, clusters are found to promote entry, but are not necessarily conducive to promoting firm growth or survival (Frenken et al., 2015). In biotechnology regional innovation systems based on research universities, spin-offs from these higher education institutions tend to have more patents than other firms, but are not necessarily more profitable or more resilient (Niosi and Banik, 2005). Thus, anchor tenants are a favourable factor in the development of a high-tech cluster, but they are neither a necessary nor a sufficient condition.

Thus, our hypotheses are formulated as follows:

Hypothesis 1: Solar clusters agglomerate around large R&D-intensive companies (the anchor tenant hypothesis in Agrawal's version).
Hypothesis 2: Solar clusters agglomerate around research universities and public R&D laboratories (the anchor tenant hypothesis in Feldman's version).
Hypothesis 3: Solar PV clusters are often semiconductor agglomerations, because solar cells – the heart of solar PV systems – are based on semiconductor-related industries (the Porter/Frenken/Boschma hypothesis).
Hypothesis 4: Solar clusters are located in large metropolitan areas where many technologies and regional innovation systems coexist (the Jacobs hypothesis).
Hypothesis 5: Solar clusters are located in specialized regions (the Marshall hypothesis).

5.2 THE SOLAR PV CLUSTERS

The solar PV sector comprises four industries. The primary one is the solar cell sector. Solar cells are specialized semiconductors transforming sunlight into electricity. The second sector is solar glass; much more recent, this highly concentrated sector has only four major players, all of them large corporations: Guardian and Corning in the United States, the British Pilkington (now a Japanese subsidiary) and the French Saint-Gobain. The third sector is composed of the providers of solar batteries – a fairly dispersed trade. Basically, solar batteries are batteries. The fourth is another dispersed sector, the one that provides metal supports for solar panels, including both metal folding and welding and mechatronics companies. The vast majority of patents are in solar cells, followed by solar glass. Mechatronics systems are also important, but while they can add efficiency to the solar panels, they are also costly and may require repair. The case of mechatronics is not yet argued. Some of the solar trackers add up to 75 per cent efficiency, but other designs only add 8 to 10 per cent output, while adding high unjustified costs. The design and costs of the solar tracker are thus vital. In sunny regions of Africa, Asia, America, Australia and Southern Europe solar systems can be beneficial without solar trackers.

Our first question was: do solar PV innovative firms agglomerate, like those in other advanced technologies, and if they do, what explains the patterns of growth, change and decline? Thus, we were interested in the dynamics of solar clusters.

Solar clusters are important because solar PV technology is positioning itself as the most likely winner in the competition between different renewable technologies that include geothermic, hydroelectric, tide and wind. The first three of these sources of energy cannot be used in many different areas of the world for natural reasons. Conversely, the rapid advance of solar photovoltaic technologies makes them increasingly efficient in many different climates and at different levels of sun exposure. Today, solar PV technologies are competitive with any other source of energy in countries with high solar exposure, such as Italy, Spain, Portugal and Greece, and large parts of China, Japan and the United States, not to mention many developing countries in Africa, Asia and Latin America. In addition, distributed energy structures minimize the cost of transmission facilities. Moreover, present-day solar technologies produce electricity during the day at the times of peak demand, while improvements in battery technologies allow its conservation during the night. Furthermore, the production of electricity close to the users' location reduces the losses of electricity on transmission lines. Finally, compared to its closer competitor, wind technology, solar PV uses fewer resources such as land and capital (US National Academy of Science Panel on electricity from renewable resources, 2010).

5.3   METHODOLOGY

In order to find out what kind of regional agglomerations, if any, could be found in solar PV technologies, we used solar patents granted by the USPTO. We found other methodologies, such as production data by cluster, basically impossible to use. Many solar PV companies are private firms, particularly in the downstream segments of the sector such as the assembling and installation of solar panels. There are no data on these manufacturing and service companies. In addition, data about industrial production of solar panels or cells are often too aggregated both in industrial and geographical terms, and therefore unusable.

We employed the US Patent Office data instead of the European Patent Office database because the United States has been the cradle of the sector between the 1950s and the 1970s, and it is still the most important country in terms of invention in its key component: solar cell patents. In the most recent segment of solar glass, only four companies in the world have obtained patents, and two of them are based in the United States. Also, only the USPTO database allows us to find the location of inventors.

We distinguished solar PV patents from other ones through keywords in the abstract, such as 'solar cell', 'solar cells', 'solar glass' and 'solar glasses'. US and international patent classifications are fairly useless, because solar cells are specialized semiconductors, and their production methods overlap with those of other semiconductors.

In order to classify the patents into different clusters, the following methodologies are employed.

5.3.1   Counting of the Number of Patents

The counting is based on the city in the address field and the counts of distinct patents. For patents with multiple inventors in multiple locations, each location will be counted once even if there are several inventors in that location. For example, for a patent with eight inventors linked to four different cities (four inventors in four different cities), it is counted once for each city. In our study, we looked at the count of patents across all the companies, and because the cities with too few patents cannot be named as a cluster, the cities with around 9 or even fewer patents will not be counted as the clusters in our study. With this counting method, it is found that 63 per cent of total patents in USPTO are not produced in the clusters.

5.3.2   Data Incomplete

As the information about the location is incomplete, we adopt the corresponding solutions:

1. Because patents with the US marked as assignee country can only be traced back to 2006, the patents in previous years only have the states marked as assignee states and without US as the assignee country, so we have to identify the assignee country as the US manually if their assignees are located in a US state. After identifying the patents before 2006, we can merge the two parts of the data together.
2. If there are several companies for which the city is missing, we have to search the company information on their website to find the correct location.
3. Both the locations of inventors and corresponding assignee names are identified. The inventors' location will be counted for the patents in certain metropolitan areas, but the patentees in that area will only include the companies as the assignees in that area.

5.3.3   From City to Metropolitan Area to Country

The information obtained was the name of individual cities, but the data have to have the metropolitan area. So we redefined and grouped the cities into different metropolitan areas, using the following criteria:

- For US: 381 metropolitan areas, http://en.wikipedia.org/wiki/List_of_Metropolitan_Statistical_Areas (Greater San Francisco, Greater Los Angeles, Greater Boston and New York are named after merging the nearby metropolitan areas).
- For Japan: 14 metropolitan areas, http://en.wikipedia.org/wiki/List_of_metropolitan_areas_in_Japan_by_population.
- For Germany: 11 metropolitan areas, http://en.wikipedia.org/wiki/Metropolitan_regions_in_Germany.
- For South Korea: 17 metropolitan areas, http://en.wikipedia.org/wiki/List_of_cities_in_South_Korea (metropolitan cities and provinces are indicated as different metropolitan areas, so a total of 17 are found).
- For Taiwan: 7 metropolitan areas, http://en.wikipedia.org/wiki/List_of_metropolitan_areas_in_Taiwan.

5.3.4 The Definition of Cluster

In total 103 metropolitan areas are identified, and only metropolitan areas with more than 42 patents (nearly 1 per cent of global patents) are defined as a cluster. In this way, there are a total of 23 clusters all over the world, which is 10 in the USA, 7 in Japan, 2 in Germany, 1 in South Korea and 3 in Taiwan.

5.4 RESULTS

5.4.1 Solar PV Clusters Worldwide

A preliminary analysis of the countries involved in solar PV technology patenting from 1976 to 2013 found that five of them represented the vast majority of the patents: the United States, Japan, Germany, Taiwan and South Korea were the leading inventors. Other Western OECD countries such as Australia, Canada, France and Switzerland also made important contributions (see Table 5.1).

We defined clusters as metropolitan regions, based on the definition and geographical delimitation used in each country. In the United States we are dealing with 'metropolitan areas'; in Japan they are 'prefectures'.

A first result was that solar PV technologies cluster like other high-tech agglomerations: the vast majority of the patents in each country were large metropolitan areas such as Greater Boston and the Greater Los Angeles metropolitan area in the United States, Tokyo and Kyoto-Osaka in Japan, Seoul in South Korea, Munich in Germany, and Taipei in Taiwan. The

Table 5.1   USPTO solar PV patents by country, 1976–2013

Country	Solar PV patents granted by the USPTO		Solar PV patents granted by the USPTO (cumulative %)
	Number	Percentage of world total	
USA	2187	48.83%	48.83%
Japan	1023	22.84%	71.67%
Germany	414	9.24%	80.91%
Taiwan	221	4.93%	85.85%
S. Korea	211	4.71%	90.56%
UK	50	1.12%	91.67%
France	40	0.89%	92.57%
Switzerland	43	0.96%	93.53%
China (incl. H. Kong)	38	0.85%	96.63%
Canada	34	0.76%	95.02%
Netherlands	34	0.76%	95.78%
Australia	33	0.74%	94.26%
Sweden	19	0.42%	97.05%
Austria	18	0.40%	97.45%
Belgium	16	0.36%	97.81%
Israel	12	0.27%	98.08%
Italy	10	0.22%	98.30%
Others	76	1.70%	100.00%
Total all countries	4479	100.00%	

only cluster that did not correspond to a metropolitan area was Silicon Valley, the largest of them all (see Table 5.2).

A second major result was that in each cluster at least one major corporation and sometimes more than one, or a large PRO, obtained the largest share of the patents (Table 5.3). Solar PV regions are 'anchored' by large firms, with one exception: in Taiwan, where the largest patentee, the anchor, is the Industrial Technology Research Institute (ITRI). In Sydney, which is a much smaller cluster, the largest innovator is the University of New South Wales School of Photovoltaic and Renewable Energy Engineering and ARC Centre of Excellence for Advanced Silicon Photovoltaics and Photonics, founded in 1975. The difference between solar clusters in Australia and Taiwan is that government-led industrialization in Taiwan did not stop after providing R&D funds to public institutions, but moved from invention to innovation to sector (Amsden and Chu, 2003), while Australia only funded academic research, and was less involved in the choice of the corporate user

Table 5.2   *Solar patents by main metropolitan area (MA), 1976–2013*

Country	Metropolitan area or prefecture	Number of solar PV patents	% of global solar PV patents	% of national solar PV patents	Total % of national solar PV patents*
USA	Greater San Francisco, CA	369	8.24%	16.87%	53.41%
	Greater Los Angeles, CA	238	5.31%	10.88%	
	Greater Boston, MA	126	2.81%	5.76%	
	Washington, DC	94	2.10%	4.30%	
	New York, NY	78	1.74%	3.57%	
	Princeton, NJ	55	1.23%	2.51%	
	Albuquerque, NM	54	1.21%	2.47%	
	Delaware Valley, DE	53	1.18%	2.42%	
	Seattle, WA	53	1.18%	2.42%	
	Dallas, TX	48	1.07%	2.19%	
Japan	Tokyo	217	4.84%	21.21%	97.56%
	Nara	211	4.71%	20.63%	
	Kanagawa	188	4.20%	18.38%	
	Kyoto	152	3.39%	14.86%	
	Osaka	107	2.39%	10.46%	
	Hyogo	75	1.67%	7.33%	
	Shiga	48	1.07%	4.69%	
Germany	Munich	122	2.72%	29.47%	39.61%
	Frankfurt/Rhine	42	0.94%	10.14%	
S. Korea	Seoul	143	3.19%	67.77%	67.77%
Taiwan	Taipei MA	74	1.65%	33.48%	74.21%
	Hsin-Chu MA	48	1.07%	21.72%	
	Taoyuan-Zhongli MA	42	0.94%	19.00%	
Total		2637	58.87%		

Note: * There is some overlapping among the different metropolitan areas due to multiple inventors for one patent.

of its technology. Thus Taiwan is among the largest patentees in solar PV technology and the second largest exporter of solar equipment; Australia has few patents and little production of this equipment.

A historical analysis of the patenting sequence in each region confirms the anchor tenant hypothesis. The large corporations (like Canon in Japan, ITRI in Taiwan and the University of New South Wales (UNSW) in Australia) were the first inventors in their respective clusters. In addition, the anchors were typically large multi-technology corporations such as Samsung and LG in Seoul (the only cluster in that country), Canon,

Table 5.3   Main patentees by regional clusters by Metropolitan Area (MA)

Country	MA	Main patentee (anchor)
USA	Greater San Francisco	Sunpower, Applied Materials, Solopower, Miasole, Solexel, Nanosolar, Solyndra, Solaria
	Greater Los Angeles	Hughes Aircraft, Atlantic Richfield, TRW, Hughes Electronics, Spectrolab, CALTEC, The Aerospace
	Greater Boston	Varian Semiconductor Equipment, Mobil Solar Energy Corp., MIT, Evergreen Solar, Spire
	Washington, DC	US governmental agencies (NASA, US Army)
	New York MA	IBM Corp. (Armonk, NY), General Electric Co., RCA Corp. Plasma Physics, Union Carbide
	Albuquerque	Emcore Solar Power, Sandia National Laboratories
	Delaware Valley	EI du Pont de Nemours, University of Delaware
	Seattle	Boeing, Allsop, Inc.
	Dallas	Texas Instruments, Exxon Mobil Corp.
Japan	Tokyo Prefecture	Mitsubishi, Showa Shell Sekiyu, Semiconductor Energy Laboratory Co., Nippon, Kabushiki Kaisha Toshiba
	Kyoto Prefecture	Canon Kabushiki Kaisha, Panasonic, Matsushita Electric Industrial, Kaneka
	Osaka Prefecture	Sharp Kabushiki Kaisha, Matsushita Electric, Sanyo Electric, Panasonic
Germany	Munich	Siemens (abandoned in 2012)
	Frankfurt/Rhine	Licentia Patent-, Merck KGaA, Nukem
S. Korea	Seoul Capital	Samsung, LG
Taiwan	Taipei-Keelung	ITRI, Atomic Energy Council – Institute of Nuclear Energy Research, National Taipei University of Technology
	Hsin-Chu	ITRI, National TsingHua University, National Chiao Tung University
	South-TW-Park	ITRI, Eternal Chemical Co., Ltd., National Kaohsiung University of Applied Sciences
	Central-TW-Park	ITRI Nexpower Technology, TSMC Solar

Kyocera, Mitsubishi and Sanyo in Japan, Telefunken, Bosch and Siemens in Germany (Granstrand et al., 1997) or multi-technology universities. In the United States, with the largest number of clusters, large firms such as Applied Materials, ARCO, Boeing, DuPont, EMCORE, IBM, Lockheed Martin, Raytheon and Spectrolab are the anchors in most clusters. Universities played a minor role in the growth of the technology, compared to large firms, with the already mentioned exception of the small UNSW cluster in Australia (Han and Niosi, 2016). In the US, Silicon Valley had a university (UCSF) and a large number of companies innovating in solar PV technologies.

Related variety also counts. In a study of the US semiconductor sector and its clusters over more than three decades, Ketelhöhn (2006) found that California, Massachusetts and New York hosted the most important US semiconductor clusters, but not the states of Delaware, New Mexico or Washington, DC. In Asia, Seoul (South Korea), Tokyo and Osaka (in Japan), Hsin-Chu and Taipei (in Taiwan) are all among the major semiconductor clusters in East Asia. Thus, out of the thirteen largest solar PV clusters, the semiconductor sector is a major employer in nine of them. Many of the large patentees in solar cells are also important ones in semiconductors (Table 5.3).

5.4.2 The Different Performances among the Different Clusters

When the patents in the different clusters are counted and the percentage of the big and small companies' patents in the total number patents is calculated (see Figures 5.1, 5.2 and 5.3), we found three categories of clusters in the world.

5.4.2.1 The resilience of US clusters

In the United States, we did not find a similar dismembering of the clusters when an anchor tenant abandoned the solar PV sector. In that country, it is easier than in Europe to form new companies; thus the closure of a large firm's laboratory or plant was most often followed by the creation of new firms in the same region. Another major element of cluster stability was its geographically established labour pool, often fuelled by university researchers. Public backing for the sector was also evident in the constant support of the Department of Energy (DOE), mainly through direct grants to innovative small and medium-sized

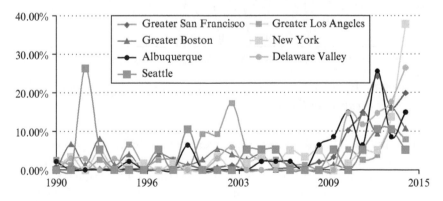

Figure 5.1   Patent distribution trends in US

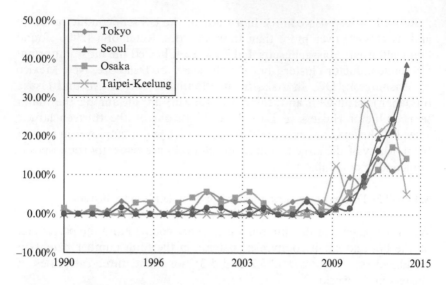

Figure 5.2   Patent distribution trends in Asia

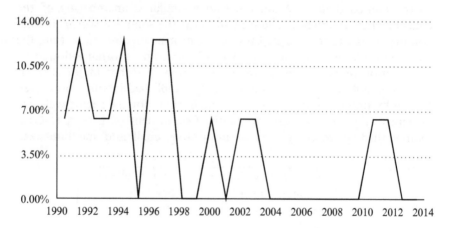

Figure 5.3   Patent distribution trends in Munich as the representative cluster in Europe

firms. In the United States, no government laboratory played such a major role as in Taiwan's ITRI, but the National Aeronautics and Space Administration (NASA) in Washington, DC, Sandia National Laboratories in Albuquerque, and several universities contributed to reinforce these anchored clusters.

Another characteristic of the US solar PV sectoral system was the fact that the main public laboratory dedicated to solar PV and other renewable energy technologies, the National Renewable Energy Laboratory (NREL), located in Golden, Colorado, a suburb of Denver, did not generate a cluster. Up to early 2015, the NREL had produced one solar PV spin-off company in that geographical area (TDA Research). NREL had over 500 patents, but just 38 US solar PV patents, and over 11 000 articles, but only around 200 solar PV articles in the Scopus database. On the Laboratory's Internet site, 12 technical success stories in different areas related to renewable energy are presented. None of them is in solar PV technology. Nevertheless, NREL is a world reference in the measurement of the efficiency of solar cells. The reason for this apparent anomaly is that US national laboratories have very different missions in the American national innovation system (Crow and Bozeman, 1998). Some of them are producing science, others have defence missions, still others are working to provide support to sectors, and among these one finds very different combinations of missions and resources. NREL provides support to the renewable energy sector, but direct promotion of economic growth is not among its missions.

Another key element in the regional agglomerations of PV solar is that venture capital has never been a major supporter of these technologies. Venture capital, mostly an American industry, has funded ICTs, biotechnology and nanotechnology, but solar PV, which seems to promise long-term returns, is far from the venture capital sector's priorities and its short-to-medium-term vision. As of today, using US National Venture Capital Association figures, we find that the funds allocated to solar PV are in downstream activities such as solar panel manufacturing and installation, and not so much invested in discovery, R&D and technical development. The fact that the United States has been at the frontier of this technology for fifty years is due to federal government support through ARPA-E,[1] DARPA,[2] SBIR,[3] the DOE[4] and other public sector organizations. Also, large US private firms in aerospace and ocean oil and gas exploration requiring a source of energy far below energy grid costs have invested in solar technologies. Thus, the only cluster where one finds dozens of small and medium-sized innovating solar PV firms is Silicon Valley, a cluster quite different from almost all others in the world in terms of entrepreneurial culture. In Germany, the federal government supported the development of solar cell technology under the aegis of Telefunken, AEG and later Siemens, for aerospace uses. Since the 1980s, German aerospace companies such as MBB have also invested in solar PV cells. But explosive growth in the German solar industries arrived only in the mid-1990s (Jacobsson et al., 2004). The feed-in tariff nurtured even further the development of a local German sector in a few clusters. With the subsequent economic crisis after 2007, the reduction of government feed-in tariff subsidy and Chinese competition, the interest of German companies began to fade away.

5.4.2.2 The growth of Asian clusters

In a sense, in its managerial practices NREL is the very opposite of Taiwan's ITRI. ITRI has been the locomotive of the industrialization of the Chinese island, having spun off dozens of companies out of its more than 6000 patents. ITRI is a multi-technology laboratory whose forte is information and communication technologies, including semiconductors, hence solar cells (Amsden and Chu, 2003). NREL has fewer patents, and therefore less technology to transfer, and seems much less eager or ready to create a cluster of spin-off companies around its facilities.

Supported by their private sector anchors, Japanese and South Korean anchored clusters have kept growing, as governments chose renewable energy – and particularly solar – as their future energy sources. The private sector anchors are large electronics firms, major producers of semiconductors.

First in Japan and later in other South East Asian countries, the production of pocket calculators, watches and other portable devices requiring a movable source of energy has impelled companies, since the 1960s, to adopt solar PV technologies. More recently, after 2010, as the efficiency of PV solar systems approaches parity grid cost, several countries in the region, namely China, Japan, South Korea and Taiwan, have been investing heavily in R&D and innovation, at the same time as the European Union countries, and particularly Germany, the European leader, are curtailing their innovation effort.

5.4.2.3 The decline of the European Union clusters

Many factors have contributed to the decline of large German and smaller European clusters. In Germany, the exit of the three largest firms (Telefunken first, and then Bosch and Siemens) triggered the decline of these inventing regions. Private sector anchors had had a large role in the growth of the clusters, but after 2013 large private anchor tenants abandoned the sector. In fact, the major threat to the continuity of the solar PV regional innovation systems, at least in Europe, is the decline of public sector support. During the economic crisis, governments reduced the feed-in tariffs that had nurtured the PV clusters (Glover, 2013). Chinese competition was also a major factor.

5.5 CONCLUSION AND POLICY IMPLICATIONS

Like other sectors based in advanced technologies, solar PV firms tend to cluster in metropolitan areas, particularly large diversified ones. The main bearers of innovation efforts are large established corporations, users of solar technologies, such as electronics firms, aerospace firms, and oil and gas companies with offshore exploration activity. These are the anchors of most clusters. The only exceptions are the Taiwanese clusters, where national laboratories (mainly ITRI) are the anchors. Last but not least is Silicon Valley, the largest PV solar regional innovation system in the world. Silicon Valley is widely seen as a class by itself among clusters. Its entrepreneurial culture, including high-tech expertise, informality and minimization of hierarchy may explain its resilience in spite of a recent period of adverse market conditions (Saxenian, 1994; Kenney, 2000). Thus, Hypothesis 1 is largely confirmed, but Hypothesis 2 is only valid for the Taiwanese clusters, due to its particular development strategy based on a large government laboratory, and publicly led industrial development (Hsu et al., 2003). Whatever the specific histories, our solar clusters are clearly anchored ones. Their anchors are most often semiconductor firms and a public laboratory only in Taiwan (as stated by Hypothesis 3).

Thus, we find organizational diversity in solar PV clusters. However, authors are divided on the issue of how much organizational diversity is conducive to cluster growth and resilience. For some (Mangematin and Errabi, 2012), organizational diversity is an obstacle to cluster growth. For others, diversity allows for the combination of more ideas and business models, in different ways, thus allowing for resilience, variety and growth (Chaminade, 1999; Agrawal et al., 2014). Our figures tend to support the Jacobs hypothesis as modified by Frenken et al. (2007) and Boschma and Iammarino (2009): large diversified metropolitan regions (Boston, Los Angeles, Munich, New York, San Francisco, Seoul, Tokyo and Osaka) are the hosts of the majority of the solar PV clusters, particularly when the anchors are active in related technologies (semiconductors). Policy implications are clear: in order to foster a solar PV cluster, governments must be sure that a variety of agencies and firms populate the cluster: large multi-technology corporations, small firms, government laboratories and research universities.

Even if after 2008 solar PV technologies have known exponential growth in terms of innovation and patents, some clusters are closing down, due to the exit of their anchor tenants. This cluster decline happened most evidently in Germany as one after the other, aerospace, automobile and electronics companies closed down their solar PV plants and R&D laboratories. The EU recession, the euro crisis, BREXIT, and

Table 5.4   Factors influencing the growth and resilience of solar PV clusters

Factors	United States	Germany	Japan, South Korea, Taiwan
Anchor tenant	Yes	Not any more	Yes
Availability of feed-in tariffs	Yes	Not any more	Yes
Collocation of semiconductor and solar cell R&D and production	Yes	Yes	Yes
Emphasis on other renewable energy industries	Yes	Not any more	Yes
Availability of venture capital	No	No	No
Status of solar PV clusters	Growing	Declining	Growing

increasingly restrictive subsidy policies continue to drive down prices and reduce EU markets. In addition, China's rapidly growing solar PV sector that now represents over 50 per cent of global solar panel production tends to depress world prices, and outcompete potential competitors, such as those in the EU, at least in terms of price. In 2013 Chinese solar panels supplied over 80 per cent of the EU market (Vega, 2013). In addition, most of the semiconductor sector is concentrated in the United States and East Asia; the EU is an increasingly marginal producer of these electronic devices.

Advanced technology clusters in the West may thus recede and disappear particularly when their stability depends on just a major multi-technology company that may retreat either from the sector or from the country (Table 5.4). Thus, a more diverse cluster, hosting research universities, large multi-technology corporations, public laboratories, SMEs and venture capital, such as Silicon Valley, may be more resilient than clusters based on one or two large firms (Table 5.5 and Figure 5.4).

In order to explain why regions differ in resilience, Martin and Sunley (2015) put forward three types of factors: compositional, collective and contextual. The contextual factors are multi-scale and they include wider conditions and forces, such as national policies and circumstances, and even international influences. After exploring the different development paths, we have identified the following factors that influence the resilience of solar PV clusters.

In a sector that has been fuelled by long-term public patent investment, the decline of public support (or its persistence) are among the main factors explaining the resilience of the sector. The clusters are anchored

Table 5.5   Large and small organizations as patentees by cluster (1976–2013)

Cluster	Large firms	Universities	PROs	Total
Greater San Francisco MA	223	86	12	321
Greater Los Angeles MA	33	37	5	75
Greater Boston MA	49	19	8	76
Greater NY MA	41	13	5	59
Albuquerque MA	42	5	1	58
Delaware Valley	29	1	5	35
Seattle MA	19	0	0	19
Munich, Germany	14	2	0	16
Tokyo Prefecture	82	41	2	125
Osaka Prefecture	44	24	1	69
Seoul Capital Area	130	15	26	168
Taipei-Keelung MA	14	12	30	56
Hsin-Chu MA	5	19	41	61

Source: USPTO.

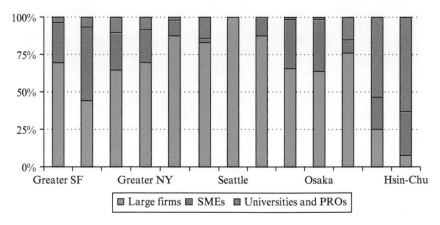

Source: USPTO.

Figure 5.4   Solar PV clusters according to main type of patentees

either by very large firms involved in semiconductors and other electronic products requiring solar energy, or by public laboratories being the arm of public policy.

NOTES

1. ARPA-E, the Advanced Research Projects Agency-Energy, is a US government agency whose mission is to fund advanced energy projects; it reports to the Department of Energy.
2. DARPA, the Defense Advanced Research Projects Agency, is an organization within the US Department of Defense whose mission is to fund R&D projects with military uses.
3. SBIR, the Small Business Innovation Research Program, is a US federal programme that intends to develop R&D activities in small and medium-sized firms.
4. The US Department of Energy (DOE) supports the development of solar technologies through different programmes, the most important being the SunShot Initiative, but also through the SBIR Programme.

6. Star scientists in PV technology and the limits of academic entrepreneurship

6.1 INTRODUCTION

Numerous scholars study the technological competencies of new technology-based spin-offs. Zucker et al. (1994) and Zucker and Darby (1996) launched a small but influential addition to this line of thought, arguing that the biotechnology revolution is mainly the result of star scientists' efforts. Zucker et al. (1994) and Zucker and Darby (1996) found that star scientists transferred their advanced knowledge to new technology firms through different channels, including participating in the scientific committees of these firms and even serving as founders or advisors. According to them, a star scientist is someone who discovers 40 or more genetic sequences and publishes at least 20 articles reporting these discoveries.

In subsequent years, some authors studying biotechnology decided to adopt this notion of star scientists and not produce original definitions (Sapsalis et al., 2006; Tzabbar and Kehoe, 2014). Others turn to patents and publications (see Table 6.1). For example, Niosi and Queenton (2010), studying Canadian biotechnology firms, define biotech superstars as those who appear as inventors in more than five patents and author of more than one major publication per year.

Although the vast majority of articles on star scientists focus on biotechnology, the concept migrated, albeit modestly so, to other high-technology disciplines such as nanotechnology, chemistry, computer and electrical engineering, and materials (Lowe and Gonzalez Brambila, 2007; Trippl, 2011; Tartari et al., 2014).

So the question for this chapter is how to define the star scientist in the solar PV sector? What are the contributions of star scientists and university spin-offs in the solar PV innovation?

Our research finds only one scholarly work on star scientists in the solar photovoltaic (PV) sector, but the author of that work (a thesis) did not build a definition (Colalat, 2009). Fuller and Rothaermel (2012) define star scientists as faculty founders of new technology ventures and apply this

Table 6.1 Star scientist definitions

Author	Sector or technology	Definitions
Zucker and Darby (1996)	Biotechnology	'Those discovering more than 40 genetic sequences and/or authoring 20 or more articles reporting such discoveries up through early 1990'
Sapsalis et al. (2006)	Biotechnology	There is no definition of stars
Lowe and Gonzalez Brambila (2007)	Six disciplines from biology and chemistry to computer and electrical engineering and materials	Stars are highly productive scholars that become entrepreneurs There is no mention of patents or the discovery of genetic sequences in the definition
Groysberg et al. (2008)	Wall Street research analysts	'... disproportionately productive and valuable' people (p. 1213)
Niosi and Queenton (2010)	Biotechnology firms and academics	Biotech superstars are those with more than five patents and more than one major publication per year
Trippl (2011)	All scientific disciplines in university	'Star scientists are defined here as authors of highly cited research papers, identified by the number of citations they generated in journals in the ISI databases in the period 1981–2002' (p. 1654)
Fuller and Rothaermel (2012)	All high-tech scientific academic disciplines	Faculty founders of new tech ventures are star scientists
Oettl (2012)	Immunology	Stars are people with high levels of scientific productivity (publications) and helpfulness. Highly productive individuals who do not help colleagues are 'lone wolfs', not 'stars'
Moretti and Wilson (2013)	Biotechnology	'... those patent assignees whose patent count over the previous ten years is in the top 5% of patent assignees nationally' (p. 3)
Tzabbar and Kehoe (2014)	Biotechnology (industrial organizations)	'Star employees have been defined as individuals who demonstrate disproportionately high levels of productivity' (p. 452). There is no definition of what high productivity may be
Hoser (2013)	Nanotechnology	Those academics with the maximum number of citations
Tartari et al. (2014)	All scientific disciplines in university	'We define star scientists as academics in our sample in the top 1% of the distribution of citations in their discipline, and the top 25% of the distribution for grants received from the EPSRC'

definition to several industries, including the photovoltaic sector. Fuller and Rothaermel (2012) mention SunPower, a photovoltaic spin-off from Stanford University in California that Dr Richard Swanson, a professor of electrical engineering founded, as a case in point. Table 6.1 summarizes the definitions and key bibliographies on star scientists.

University spin-offs (USOs) are one of the important channels through which star scientists can contribute to the growth of high-tech firms. Pirnay et al. (2000) define USOs as follows: 'new firms created to exploit commercially some knowledge, technology or research results developed within a university'.

The scientific domain itself makes sense for the performance of USOs. A large percentage of academic spin-offs relate to biotechnology and health sciences. Mowery et al. (2001) calculate that some 75 per cent of the patenting and licensing at three of the most research-active universities in the United States, namely, the network of the University of California, Columbia and Stanford, are in biomedical research, particularly biotechnology. The second most important sector is computer software. Mowery et al. (2001) does not mention solar technology. Similarly, in an annual survey of intellectual property generated in Canadian universities, health sciences appear as number one, although not as prominently as in the US. Again, the survey does not mention the solar photovoltaic (PV) sector (Statistics Canada, annual).

Local venture capital also appears to be a determinant of growth. Using a large sample of US academic spin-offs supported by venture capital, Zhang (2009) finds that most of them focus on two areas: biotechnology and information technologies. In addition these spin-offs tend to remain geographically close to their alma mater.

USOs from different US universities perform very differently. More entrepreneurial universities have much better scores as licensors of technology to academic spin-offs. Walter et al. (2006) argue that network capabilities and entrepreneurial orientation are key variables explaining the performance of these USOs. Further, Powers and McDougall (2005) find that universities with experienced (older) technology transfer offices (TTOs) incubate more successful spin-offs. More productive faculty members (in terms of articles and citations) also contribute to more successful spin-offs. Early collaboration with the sector also contributes to spin-off growth.

6.2 HYPOTHESES

On the basis of literature reviews, the chapter draws the following hypotheses:

Hypothesis 1: Technological content (that is, the relevant industrial sector) is a major determinant of the likelihood of the creation of a USO. More specifically, the likelihood of creation of a solar PV USO is lower than that of biomedical and information technology USOs.

Hypothesis 2: Venture capital has a strong industrial sector component. Venture capital supports biomedical and information and communication technologies (ICT) spin-offs much more often than solar PV ones.

Hypothesis 3: More experienced universities and TTOs will produce more successful USOs.

Hypothesis 4: Star scientists will engage in more successful solar PV USOs.

6.3   THE SOLAR PV SECTOR

The solar photovoltaic sector started modestly in Bell Labs in the 1950s when three researchers developed the silicon transistor. Today, the silicon cell is still the main component of solar PV technology (Perlin, 2002). The first application was in space, in the late 1950s and 1960s, with satellites requiring a reliable long-term source of electricity, even if the cost of that solar energy was high. A second major application, in the 1970s, was in sea buoys and sea oil and gas exploration and exploitation far from conventional sources of electricity. At this time, large hydrocarbon companies, such as ARCO, BP and Shell, started investing in solar PV R&D. As the cost of solar PV energy started to decline following technological advances, the sector began to interest companies such as Telecom Australia, as a means to provide telephone connections in a country close to 8 million square km in area and with lots of sunshine but with a population of only 12 million in the early 1970s. At the same time, Japanese companies such as Canon, Sharp and Sanyo invested in solar technology to power their hand-held calculators and similar devices.

Universities joined the solar bandwagon later. In the 1980s, the University of New South Wales (UNSW), under the guidance of Dr Martin Green, started conducting research on solar cells to improve their efficiency. At the same time, Dr Allen Bennett at the University of Delaware (UD) provided an impetus to academic research on PV technologies. Soon, these two pioneers launched the first academic spin-offs, Pacific Solar in Sydney Australia and AstroPower in Glasgow, Delaware, in 1995. AstroPower sold its assets to GE in 2004, and GE handed these assets to Taiwan's Motech in 2009. The main UNSW spin-off, Pacific Solar, experienced ups and downs and finally closed its doors in the late 1990s. Several other universities, not only in the United States but also in Europe (Germany and Switzerland), had started to conduct R&D on solar PV technologies in a multi-agent race to increase solar cell efficiencies by that time. Today, the University of Konstanz in Germany and the École Polytechnique Fédérale de Lausanne (EPFL) in Switzerland stand among the top contributors. Star Professor Michael Graetzel at EPFL is the Chairman of the Technical Advisory Board at an Australian spin-off company, Dyesol, which acquired one of EPFL's spin-off companies, GreatCell, a firm that made use of Dr Graetzel's discoveries in the field of dye solar cells.

Germany has lost some interest in the solar PV sector due to a reduction in government feed-in tariffs and Chinese competition. In 2009, the largest German start-up, Solon Technologies, went bankrupt, and soon the two largest German companies involved in the area, Robert Bosch and Siemens, sold or closed their solar PV facilities. However, government R&D laboratories remained active, with the Fraunhofer Institute in Freiburg being Europe's number one public research institute in the area. The Fraunhofer Institute created close to 15 solar PV spin-off companies, some of which are still active.

Since 2000, Asian competitors have become more involved. Japan has developed a policy to foster the creation of university spin-offs, a phenomenon that seldom occurs in that country. However, as of 2015, there are few spin-offs from Japanese universities in the solar PV sector. All of them are working on niche products.

In Taiwan, the Industrial Technology Research Institute's (ITRI) policy is to import and develop advanced technology and spin-off technology-based firms concentrating on solar PV, after focusing on semiconductor products and processes. DelSolar, located in the Hsinchu Science Park, is ITRI's first and only PV spin-off. In 2005 DelSolar merged with NSP, another Taiwanese company, to become the largest solar PV technology firm in Taiwan.

Whereas South Korea seems almost impenetrable to solar academic spin-offs, China's Academy of Science and some Chinese universities are very active; TsingHua University stands out over the others.

Asian and US governments are increasing their investment in solar PV R&D, but venture capital (VC) is historically fairly reluctant to engage in a sector that promises returns only in the long term and is plagued by high volatility. Only in the last few years did some VC investment return to the sector, mainly in the United States, and mostly in the downstream segments of the sector, such as companies that assemble and install rooftop modules. Upstream companies, with strong R&D capabilities and patents, in which star scientists are usually involved, are struggling to obtain private sector funds and only survive due to public funds, such as those from the US Department of Energy, the Advanced Research Project Agency

(ARPA), and the Small Business Innovation Research Program (SBIR) programmes. The decline in the price of oil and gas since late 2014 is doing nothing to reverse the trend.

6.4 METHODOLOGY

This work focuses mainly on patents, publications and venture capital, as well as on the construction of lists of new spin-offs and start-ups in the solar cell sector.

The authors use the United States Patent and Trademark Office (USPTO) database and the keywords 'solar cells', 'solar cell', 'solar glass', 'photovoltaic cells' and 'photovoltaic cell'. Removing the overlaps, the authors find some 4400 solar PV patents granted between 1976 and December 2013. The use of the USPTO database is because close to 50 per cent of the solar PV patents originate in the United States and competitors in Japan, Germany, Taiwan and the People's Republic of China also apply for patents in the United States to protect their inventions from potential infringers.

Following the most widely used definition of a star scientist, the authors first analyse the articles of star scientists. There are 105 484 articles on solar technologies in Scopus as of 17 February 2015. Within this group, there are 109 stars with more than 100 articles with 'solar cell' as the search query for the 'Article title, abstract, and keywords' (see Table 6.3).

For venture capital, the authors use the secondary Venture Capital Association reports as proof.

6.5 RESULTS: STAR SCIENTISTS IN THE PHOTOVOLTAIC INDUSTRIES

6.5.1 Definition of Star Scientist for Solar PV Sector

Analysing the correlation between the number of articles in Scopus and the number of patents by academic scientists in USPTO and Espacenet, the correlation coefficients are just 0.043 and 0.22, respectively. The reason for this high-tech 'anomaly' is that at least in North America, solar technology receives enormous knowledge externalities from several other industries, including semiconductor firms and large glass producers investing in the speciality technical glass used in PV equipment. Thus, for the solar PV sector, unlike biotechnology or other disciplines, most of the research activity takes place in firms, with rather less in universities.

For photovoltaic cells, universities have just 589 US patents, whereas companies have more than 3800. There are almost 41 800 articles in Scopus on photovoltaic cells. Just over 55 per cent of them include authors from universities. For photovoltaic glass, of the 3600 articles in Scopus, some 2073 (or 57 per cent) include academics among the authors. In addition, photovoltaic glass requires large manufacturing plants and R&D activity. The four main patentees are very large American, European and Japanese companies (Guardian Industries and Corning, both from the United States, the French Saint-Gobain and Japanese Nippon Sheet Glass, with its British Pilkington subsidiary). In Germany, some universities conduct solar technology research, particularly Albert Ludwig, Konstanz and Martin Luther universities. However, the number of German academic patents is very low. Only seven appear in the USPTO database, all granted after 2004. The assignees are the Universities of Konstanz and Albert Ludwig Freiburg. A few other academic patents appear in Espacenet, making a total of 15.

To reflect academic expertise, our definition utilizes articles and patents, and our definition is as follows:

In the solar photovoltaic sector, academic star scientists are university or institutional researchers with at least four photovoltaic sector patents in the USPTO and/or over 100 Scopus publications.

The cutting point of the patents and the publications are based on the following two criteria with the statistics of the patents and publication in Table 6.2.

- 80/20 principle: The total number of papers/patents is respectively nearly 80 per cent of the papers/patents that all the scientists in the university have made.
- The star scientists cannot be too many. So if there is a big difference between the consecutive number of papers or patents, the upper number will be taken as the standard of the cutting point to define

Table 6.2   Statistics for the cutting point of the criteria for star scientists

No. of patents per scientist	Total no. of patents	No. of scientists	Total patents	% of total	No. of papers per scientist	Total no. of papers	Total papers	% of total
4+	273	15	589	46.35%	100+	16 896	22 990	73.33%
3+	429	52		72.84%	90+	18 481		83.94%
2+		158						

the star scientist. So the cutting point is based on the distribution of papers/patents.

The authors with more than 200 documents in Scopus are listed in Table 6.3.

6.5.2   The Difference between Solar PV and other High-Tech Disciplines

The solar PV sector produces far fewer patents than the biomedical and ICT industries. A simple perusal of the Organisation for Economic Co-operation and Development (OECD) patents database shows that, between 1976 and 2011, OECD countries requested some 175 720 Patent Cooperation Treaty (PCT) priority patents in biotechnology, compared to 13 984 in solar PV technology. The difference, which is more than 12 times, is staggering, even when adding approximately 500 patents from China, Singapore and Taiwan not granted in the USA.

Academic patents in solar PV technologies are fairly scarce. There are 589 university or institute patents. Removing overlaps, this amounts to barely more than 10 per cent of the 4400 solar cell patents. In comparison, there are some 90 000 USPTO patents granted to assignees with the word 'university'. One half of 1 per cent of them are solar PV patents.

6.5.3   Venture Capital in Solar PV Technologies

In the USA, which hosts 50 per cent of the world's venture capital, bio-medical and ICT technologies relegate solar PV technologies to the bottom of the list by a wide margin. The US National Venture Capital Association (NVCA) put software at the top, with 41 per cent of the total investment of US$48 billion, followed by biomedical and life sciences, with 18 per cent, and other sectors in 2014. The reports do not even mention energy or clean energy, let alone solar PV technology. A Mercom Capital Group report (2015) estimated the investments in solar PV technology, putting the global total at US$1.3 billion for venture capital (most of this amount goes to downstream activities such as solar roof panel installation for residences in the United States, corresponding to two companies) and US$26.5 billion for corporate investment. In 2013, global investments for solar PV were only US$643 million from US$9.6 billion for clean technology. It is concluded that governments should support this emerging technology which will lead to global improvements in climate and environment in the medium and long term. Private venture capital is not interested in nurturing a clean environment.

Table 6.3   The authors with more than 200 documents in Scopus

Authors from search of 'solar cell' in Scopus	Documents with 'solar cell'	Country	Organization	Total documents	Total citations
Gätzel, M.	518	Switzerland	École Polytechnique Fédérale de Lausanne	993	131061
Green, M.A.	444	Australia	UNSW	639	19015
Poortmans, J.	295	Belgium	Universiteit Hasselt, Faculty of Science, Diepenbeek	408	4622
Konagai, M.	272	Japan	Tokyo Institute of Technology	514	5295
Nazeeruddin, M.K.	271	Switzerland	École Polytechnique Fédérale de Lausanne	386	37927
Schock, H.W.	266	Germany	Helmholtz-Zentrum Berlin für Materialien und Energie (HZB), Berlin	340	8728
Zhao, Y.	247	China	Ministry of Agriculture of the People's Republic of China, Key Laboratory of Plant Nutrition and the Agri-environment in Northwest China, Beijing	1410	26226
Hagfeldt, Anders	242	Sweden	Uppsala Universitet, Department of Chemistry-Ångström, Uppsala	323	26211
Schropp, R.E.I.	235	Netherlands	Technische Universiteit Eindhoven, Department of Applied Physics, Eindhoven	388	5021
Li, Yongfang	233	China	Beijing National Laboratory for Molecular Sciences, Institute of Chemistry, Zhongguancun	569	18437
Rau, U.	231	Germany	Forschungszentrum Jülich (FZJ), Jülich	281	5960
Luque-Lopez, Antonio	230	Spain	Escuela Técnica Superior de Ingenieros de Telecomunicación, Madrid	442	8874
Ballif, C.	226	Switzerland	École Polytechnique Fédérale de Lausanne	304	5170
Krebs, F.C.	217	Denmark	Danmarks TekniskeUniversitet, Department of Energy Conversion and Storage, Lyngby	365	18928

6.5.4   University Patents

The US universities and public laboratories dominate in terms of academic solar technology patents, with 292 patents. MIT, the University of Delaware, North Carolina State University, the University of California, Caltech, Princeton and Stanford appear at the top of the list.

In Taiwan, where assignees have 136 US patents, ITRI tops the list with 57, followed by the Atomic Energy Council (22) and National TsingHua University (13).

In South Korea, Dongguk University, the Korean Institute for Science and Technology (KIST), the Korean Research Institute for Science and Technology and Sungkyunkwan University own more than 50 per cent of the 82 public sector solar PV patents.

There are few academic patents in other countries. Australia would be a contender, but UNSW has only two patents under its name. Private firms and the university's intellectual property arm presented several of its innovations. China (9), Canada (8), Germany (7), and a few other countries follow. The distribution of academic papers is much greater in other technologies than in solar PV.

6.5.5   Star Scientist Performance

Stars in the solar PV sectors do not focus solely on PV technology; nearly half of their academic work is in other related fields. The average percentage of articles with the title or abstract containing 'solar cell' is just 56 per cent of their total papers, and the correlation coefficient between the number of articles with the title or abstract containing 'solar cell' and the total number of articles is 0.32. For star scientists, the correlation coefficient between the percentage of papers with 'solar cell' and the total number of patents in USPTO and Espacenet is 0.051 and 0.022, respectively, which indicates that the stars with more papers are not very active in solar PV entrepreneurship.

Among the 109 star scientists, the authors remove nine employed by companies because this work focuses on academic entrepreneurship. This procedure leaves just 100 stars belonging to 22 countries (see Table 6.4). Stars in different countries have different forms of entrepreneurship. Most of the stars outside of the US are not high on USPTO patent lists but have several patents in their own countries, especially the stars in Asian nations, such as Japan and China. Among the 22 stars in the US universities, eight of them have experience setting up a business or working experience in companies, with their academic expertise directly serving in the commercialization of the technology. In contrast, in Japan, finding

Table 6.4   *The number of academic star scientists in terms of solar PV papers in different countries*

Country	Number of stars	Location	Universities
Germany	17	Breisgau (5)	Fraunhofer-Institut für Solare Energie Systeme (5)
		Berlin (4)	Helmholtz-Zentrum Berlin für Materialien und Energie (4)
		Jülich (3)	Forschungszentrum Jülich (3)
		Hannover (2)	Universität Hannover (2)
		Dresden (1)	Technische Universität Dresden (1)
		Erlangen (1)	Friedrich-Alexander-Universität Erlangen-Nürnberg (1)
		Stuttgart (1)	Institute for Photovoltaics (1)
United States	15	CA (4)	Caltech (1), UCSB (1), Stanford (1), UCLA (1)
		CO (4)	NREL (3), Colorado State University (1)
		DE (1)	University of Delaware (1)
		OH (2)	University of Toledo (2)
		MD (1)	University of Maryland (1)
		NY (1)	State University of New York at Buffalo (1)
		PA (1)	Pennsylvania State University (1)
		Washington, DC (1)	Naval Research Laboratory (1)
Japan	15	Tokyo (5)	Tokyo Institute of Technology (2), National Institute of Advanced Industrial Science and Technology (2), University of Tokyo (1)
		Kyoto (3)	Ritsumeikan University (3)
		Kitakyushu (2)	Kyushu Institute of Technology (2)
		Tsukuba (2)	National Institute for Materials Science Tsukuba (2)
		Chofu (1)	Japan Aerospace Exploration Agency (1)
		Kawaguchi (1)	Japan Science and Technology Agency (1)
		Nagoya (1)	Nagoya Institute of Technology (1)
China	12	Beijing (3)	Key Laboratory of Plant Nutrition and the Agri-environment in Northwest China of Ministry of Agriculture (1), Institute of Chemistry CAS (2)
		Quanzhou (3)	Huaqiao University (3)
		Guangzhou (2)	Sun Yat-Sen University (1), South China University of Technology (1)
		Chengdu (1)	University of Electronic Science and Technology of China (1)
		Hefei (1)	Hefei Institutes of Physical Sciences, CAS (1)
		Shanghai (1)	Shanghai Institute of Technology (1)
		Tianjin (1)	Nankai University (1)
Switzerland	7	Dubendorf (1)	Forschung Institution für Materialwissenschaften Und Technologie Eth-Bereichs (1)
		Lausanne (6)	École Polytechnique Fédérale de Lausanne (6)
		Stockholm (1)	The Royal Institute of Technology (1)

Table 6.4   (continued)

Country	Number of stars	Location	Universities
Netherlands	4	Eindhoven (2)	Technische Universität Eindhoven (2)
		Delft (1)	Delft University of Technology (1)
		Utrecht (1)	Debye Institute (1)
Spain	4	Madrid (2)	Escuela Técnica Superior de Ingenieros de Telecomunicación (1)
		Madrid (1)	Universidad Politécnica de Madrid (1)
		Tarragona (1)	Instituto Catalán de Investigación Química (1)
South Korea	3	Jongno-gu (2)	Sungkyunkwan University (2)
		Seoul (1)	Korea University (1)
Australia	3	Sydney (2)	UNSW (2)
		Canberra (1)	Australian National University (1)
Sweden	3	Uppsala (2)	Uppsala Universität (1), Angstrom Laboratory (1)
Taiwan	3	Taipei (1)	National Taiwan University (1)
		Hsin-chu (2)	National Chiao Tung University Taiwan (2)
United Kingdom	3	London (2)	Imperial College London (2)
		Oxford (1)	University of Oxford (1)
Belgium	2	Diepenbeek (1)	Universiteit Hasselt (1)
		Leuven (1)	Katholieke Universität (1)
Denmark	1	Lyngby (1)	Danmarks Tekniske Universitet (1)
Austria	1	Linz (1)	Johannes Kepler Universität Linz (1)
Singapore	1	Singapore City (1)	National University of Singapore (1)
France	1	Paris (1)	EDF Institut de Recherche et Développement sur l'Énergie Photovoltaïque (1)
Malaysia	1	Bangi (1)	Universiti Kebangsaan Malaysia (1)
Ethiopia	1	Addis Ababa (1)	Addis Ababa University (1)
Slovenia	1	Ljubljana (1)	University of Ljubljana (1)
Israel	1	Rehovot (1)	Weizmann Institute of Science Israel (1)
Saudi Arabia	1	Jeddah (1)	King Abdulaziz University (1)

Source: Scopus.

entrepreneurial activity among stars is difficult, except in terms of local patenting. Considering that very large companies dominate the PV sector in Japan, large companies employ most of the academic Japanese stars in one way or another, and stars are less engaged in entrepreneurial activity by themselves.

6.6   CONCLUSIONS AND POLICY IMPLICATIONS

Star academic entrepreneurship and relative venture capital are most active in biotechnology, other human health sectors and ICT, including software. Solar PV is another story. Since the inception of the technology in the 1950s and 1960s, several factors have restricted the creation of solar USOs, including the scarcity of research funds (in comparison with biomedical technologies), niche markets and the modest interest of academic researchers in the subject, with only a few active universities outside the US, such as UNSW in Australia and EPFL in Switzerland. Comparatively, there are few university patents on solar PV technologies. In addition, venture capital is fairly reluctant to invest in the field, except in Silicon Valley. New solar PV technologies do not attract much interest, and there are few start-ups. Thus, the authors find moderate evidence of support for Hypotheses 1 and 2: both academia and venture capital privilege life sciences and ICT at the expense of renewable energy technologies. Only a few countries, such as Australia, China, Japan, Taiwan and the US, are fuelling innovation in this sector, most often with public monies.

Being comparatively new, the PV sector requires the accumulation of knowledge in related fields, such as semiconductors and glass technologies, advanced batteries and mechatronics. Under these conditions, scientists have greater potential to become star scientists in life sciences, ICT and nanotechnologies than in PV technologies. Today, the distribution of solar PV star scientists is global, but they are seldom entrepreneurs.

The authors cannot accept Hypotheses 3 and 4. Universities surrounded with venture capital (mainly Silicon Valley but also Greater Boston and Los Angeles) may produce more successful spin-offs (see Table 6.5). Conversely, prestigious academic institutions in the area of solar PV technology, such as EPFL and UNSW, do not produce similar numbers and successful spin-offs. The findings do not justify extending the idea that successful stars engage in more successful spin-offs in the field of solar PV technologies.

Academic entrepreneurship is not widespread in the PV sector, even in US universities where academic entrepreneurship in bio- and nanotechnologies and ICTs is very active. Furthermore, most of the successful firms

Table 6.5   Solar PV academic patents by country, state and university

Country	State	University	Number of USPTO solar PV patents
USA	Total		292
	MA	MIT	42
	DE	University of Delaware	29
	NC	North Carolina State University	22
	CA	University of California	21
	MI	Midwest Research Institute	21
	CA	California Institute of Technology	13
Taiwan	Total		136
		ITRI	57
		Atomic Energy Council	22
		National TsingHua University	13
South Korea	Total		82
		Dongguk University	17
		KIST	13
		KRICT	10
Switzerland	Total		10
		École Polytechnique Fédérale de Lausanne	10
China	Total		9
		TsingHua University	7
Canada	Total		8
	ON	University of Toronto	3
Germany	Total		26
		University of Konstanz	5
		Fraunhofer Institute	19
All other countries			45
Total			608

Source: USPTO.

that are established directly or use knowledge produced by star PV scientists in PV are being acquired or shut down. To make the scientific and technological achievements of star scientists successfully commercialized, the corresponding procedures for technological transfer between academic units and firms should be designed and operated well. Venture capital and a good regional economic landscape may also be necessary conditions for the development of successful new ventures.

From a theoretical point of view, we argue that the concept of a star scientist has to be integrated in the resource-based and competence theories of the firm to which the star scientist approach belongs. In many high-tech companies, star scientists are one of the key resources for growth.

From a public policy point of view, the overwhelming presence of large firms compared to spin-offs from academic institutions makes us think, similarly to Mowery et al. (2001), that academic entrepreneurship applies to a reduced set of technological domains. Solar PV technologies are not a central part of this set. The conditions of these technologies, namely their high risk, long-term pay-offs and strong competition from huge corporations, may cause any policy aiming to create academic spin-offs to be doomed from the start.

7.   A sector with innovations driven by demand

7.1   HOW TO EXPLAIN THE DISTINCTIVE FEATURES OF THE SECTOR

Based on the above research on the important aspects of the solar PV sector, some distinctive characteristics have been drawn:

- The majority of the innovators are industrial user firms, but not the pure solar PV firms whose primary business is solar PV. Government subsidy policies to promote usage greatly influence industrial development and innovation behaviours.
- The different generations of technologies have been coexisting for a long time. Although the new technologies have been massively exploited in recent years, the technologies with the highest market penetration rate are not the most advanced. They are cheaper, which is why they are the most diffused.
- For solar PV innovation, even the star scientists only put part of their expertise into the sector, and the rest into other related areas. This can be seen through their patents and publications. In addition, their academic entrepreneurship is quite limited.

How can we explain the above phenomenon, which is in some way different from the typical high-tech sector? The solar PV sector is not a classic high-tech sector, where innovation and industrial development are driven by scientific or technological progress. But if taken as a traditional sector, its technology developed so rapidly and the technological innovation producing cost-efficiency improved so significantly that the use of solar energy has become more and more efficient in combination with other sources of energy.

7.2 SIMILARITY WITH THE SEMICONDUCTOR SECTOR

The case of the solar PV sector is not unique. The semiconductor sector, which is the key origin sector for the solar PV sector, has the same characteristics as those of the solar PV sector in some areas.

By examining patents, co-patents, R&D alliances and new ventures in semiconductors, Adams et al. (2013) have drawn quite similar conclusions for the semiconductor sector:

1. The magnitude of innovation by user firms was quite high in both absolute and relative terms compared to firms in the sector over the entire period under examination, and a broad range of intermediate users were a major source of patents in a product field (semiconductor devices) outside of their 'core' business.
2. The distribution of innovation among firms from different intermediate user industries was highly uneven; this finding points to differences across final demand groups in terms of the requirements, the intensity of use, and the strategic content of semiconductors.
3. Innovative users were highly heterogeneous in terms of size, diversification and vertical integration. Large user firms, most of which were vertically integrated, have substantial patent portfolios. Their main line of business is not semiconductors but they produce chips as vertically integrated user firms rather than as diversified semiconductor firms. There is also evidence of a vast number of smaller user firms that were able to patent this technology, albeit at lower rates.

7.3 USER-INNOVATION SECTOR WHOSE INNOVATIONS ARE DRIVEN BY DEMAND?

According to Adams et al. (2013), the various streams of research have shown that user firms may contribute to innovation in a variety of ways. 'Active' users may simply provide knowledge and feedback to producers (Eurostat, 2004) while 'lead' users (Von Hippel, 1986; Gault and Von Hippel, 2009) will innovate on their own in order to develop solutions for their specific needs before the bulk of the marketplace even recognizes the same need. 'Experimental' users (Malerba et al., 2007) are willing to try emerging technologies and attribute intrinsic merit to a product simply because it embodies a new technology. 'User entrepreneurs' go further to take responsibility for the production and commercialization of products/ services that they have first developed for their own use (Hienerth, 2006;

Shah and Tripsas, 2007). 'Vertically integrated' user firms design and produce components for their in-house use and often sell their component solutions to the open market as well. For the semiconductor sector, 'vertically integrated' user firms ('user firms' hereafter) are the important innovators.

Adams et al. (2013) classify the actors of the semiconductor sector in five categories: Semiconductor Firms; User Firms; Academics and Professionals; Linked Industries; and Other Industries. The User Firms category consists of companies that sell products or services that use or incorporate semiconductors in six industries including: Industrial Machinery; Consumer Electronics; Computer Equipment; Telecommunications; Automotive; Instrumentation and Aerospace/Defense.

By separating the related manufacturing firms into two groups, we can classify two categories for the solar PV sector: one is composed of user firms and the other is made of specialized firms.

- User firms are companies whose main business is not solar PV products but they innovate in solar PV in order to serve their main business. Their main demand is from their specific usage (for example, to be used in extreme environments such as in space or off-shore exploration), and the demand-driven innovations are mainly conducted inside the firms.
- Specialized firms are companies whose main business is the manufacturing and sale of solar PV products. Their main demand is from daily electricity usage, in which the priority will be the low cost of the up-front installation and adequate supply of energy for daily use.

Specialized firms appeared around thirty years ago, which may seem very late compared to the sixty-year technology development history of user firms. Actually, user firms have driven technology innovation for the solar PV sector in the long run. Historically, the various applications of solar PV technologies evolved in the following order: its first application was in space in the late 1950s and 1960s, with satellites requiring a reliable long-term source of electricity, even if the cost of this energy was high. A second major application, in the 1970s, was in sea buoys and sea oil and gas exploration and exploitation far from conventional sources of electricity. At this time, large hydrocarbon companies such as ARCO, BP and Shell started investing in solar PV R&D. As the cost of solar PV energy started to fall, following technical advances and economies of scale, some companies such as Telecom Australia became interested in the sector to provide telephone connections in a country with close to 8 million square km, lots of sunshine, but with a population of only 12 million in the early

1970s. At the same time Japanese companies such as Sharp, and then Sanyo, invested in solar technology to equip their hand-held calculators and similar devices.

The number of patents owned by the top ten user firms and the top ten specialized firms are respectively listed in Tables 3.10 and 3.11. The top ten user firms hold nearly 20 per cent of total USPTO patents in solar PV while the top ten specialized firms only hold just under 5 per cent.

By separating the firms into user firms and specialized firms, we can draw almost the same conclusion as that for the semiconductor sector: the solar PV sector is a user-innovation sector whose innovations are driven by demand.

Considering the characteristic of user-innovation, the distinctive features defined at the beginning of the chapter can be explained well here:

1. Distinct academic innovation behaviour:

   - nearly half of the publications of star scientists in the solar PV sector are in the other related domains;
   - academic entrepreneurship is quite limited;
   - it is unusual that the star scientists who have a great number of publications do not have many patents, and the inventors who have large numbers of patents do not have many publications, while the star scientists in bio-industries always have both publications and patents.

   Considering the fact that the intermediate user firms are dominating the sector, the majority of the innovation is made inside big user firms, the diversified research direction will be made inside the firms and the research budget will be dispersed among the different user firms, the attractiveness of and support for the academic scientists are much less than for other high-tech industries. This can well explain why there is different innovation behaviour in the sector.

2. The coexistence of the different generations of solar technologies while the first generation still holds the biggest market share.

   There are three generations of technologies available now, but the technology with the highest marketing penetration rate is still the first generation, with which Chinese manufacturers can achieve the best performance in the world. Why did the new technology emerge so fast, while the dominating technologies are still of the first generation? The answer is that innovation is driven by demand. On the one hand, the diversified demands from the user firms drive continuous innovation; on the other hand, the massive market demands of daily electricity usage have not yet been explored. The more mature the technology is, the less costly is the production using this technology. Such technological maturity requires less and less R&D investment. Nowadays, the technologies in the first generation still have cost advantages, and it is still the best choice for the much bigger daily electricity usage market. Only on the condition that the more advanced technologies can be installed and operated with cost advantages can new generations of technology achieve higher market shares.

3. The catching-up capability of developing countries is supported by the demand for usage in low-cost technology; only countries with fast-moving integrating production capabilities can satisfy the demands of the market with lower costs. This can explain why Chinese manufacturers have been able to catch up on production and not on cutting-edge innovation and maintain the highest market share in recent years.

7.4 CAN THE SOLAR PV SECTOR BE CONSIDERED TO BE A NEW SUB-CATEGORY IN THE HIGH-TECH INDUSTRIES?

Based on the above studies, our conclusion is that the solar PV sector is not a traditional high-tech sector but a demand-driven one. Because of this significant feature, the sector has embodied a different kind of academic innovation behaviour in the evolution of the technologies and in the comparative advantages among the different countries. So the question will be: can the solar PV sector be considered to be a new sub-category in the high-tech sector? There are two concerns blurring the answers:

1. The sector is in such an early stage that it is not possible to draw any conclusions at the moment.
2. In some segments of the high-tech industries, innovation and industrial development are mostly driven by demand instead of by technological progress itself.

8. Grand challenges and innovation cascades in the solar sector

Before finishing the book, we want to put forward a new direction for the solar PV sector in the future, that is, linking two recent concepts about innovation: those of grand challenges and innovation cascades. At the same time we show how our ideas about innovation and innovation policy have evolved.

Two sections will be included in the chapter: (1) Innovation from serendipity to cascades; and (2) Innovation policy from support to routine innovation to support to cascades.

8.1 INNOVATION FROM SERENDIPITY TO CASCADES

8.1.1 The Evolution of the Concept of the Innovation Cascade

Up to the mid-twentieth century, innovation was supposed to be the result of the genius of individual scientists, inventors and engineers. From Galileo to Louis Pasteur and Thomas Alva Edison, the number of biographies devoted to these great men of industry and science has been enormous and occupies many sections of libraries. It is probably true that before the development of the modern research university and the public research laboratory, in the nineteenth century the development of science, technology and innovation fell upon the shoulders of talented and devoted individuals. In the creation of the concepts and the discovery of new laws in nature and society, there was an enormous amount of effort and some degree of serendipity. The discovery of penicillin by Alexander Fleming in the early 1930s is often mentioned as an emblematic case of serendipity.

Joseph Schumpeter was among the earlier social scientists who put innovation at the centre of his work. In his first book, *The Theory of Economic Development* (1916), he put his focus on innovation, and he linked innovation to economic development. He left invention in the shade: whoever did the inventions, he or she passed the baton to the innovator, the person who saw the potential economic impact of the novelty and who was able to translate it into products, processes and new forms of organization. In his second major book, *Business Cycles* (1939), he moved forward and hypothesized that innovation occurred in rapid succession, as other innovators that often improved the original product, process or organization imitated the first innovator. Thus a radical novelty was followed by a series of smaller improvements, that the literature has called 'incremental innovation'. The invention of the motorcar went through similar stages. In 1883, Karl Benz produced the first internal combustion engine powered car. In the late 1880s, G. Daimler and W. Maybach developed another car in Germany, while E. Bernardi developed his own model in Italy. Soon after, automobiles were designed and produced in Britain, Canada, France, Germany, Italy and the United States. Each new manufacturer brought forward it own novelties in brakes, engines or wheels, and by the early 1900s, the automobile industry was taking off. Thus the serial innovation process that Schumpeter discovered was international in scope, and was often the cradle of an industry; in this case, the auto industry.

Innovation cascades are akin to these series of innovations that Schumpeter was the first to discover. Both are international and have profound impacts on the industrial structure. The difference is that Schumpeter's series of novelties were based on one radical innovation, followed by hundreds of minor improvements. In David Lane's conception (2016), innovation cascades correspond very much to this type of serial innovation. He attributes this kind of process to a positive feedback effect where:

> (1) new artefact types are designed to achieve some particular attribution of functionality; (2) organizational transformations are constructed to proliferate the use of tokens of the new type; (3) novel patterns of human interaction emerge around these artefacts in use; (4) new attributions of functionality are generated to describe that the participants in these interactions are obtaining or might obtain from them; (5 = 1) new artefacts are designed to instantiate the new attributed functionality. (Lane, 2012, p. 1)

Instead, in our work, *innovation cascades are series of radical innovations*. They occur in different but related industries, such as – in solar – new semiconductors, new solar glass, more advanced batteries, and mechatronics. We argue that Lane's view of innovation cascades – as his example of the printing press in Italy shows – corresponds to patterns found throughout the fifteenth to the nineteenth centuries. Instead, our series of radical innovations corresponds more often to our modern high-tech industry. The reasons are many.

1.   During the last decades of the twentieth century, the number of innovating countries and organizations has increased enormously. From

the fifteenth century to the twentieth century, the vast majority of innovation was created in fewer than 20 countries, located in Europe and North America: in Britain, France, Germany, Italy, the Western, Eastern and Nordic countries in Europe, as well as Canada and the United States in North America. Today, one must add several Asian countries including China, Japan, Singapore and South Korea. Within those countries, the number of innovative firms, research universities and government laboratories has undergone a phenomenal increase.

2. Knowledge is produced and transferred much faster today than it did half a century ago, due to the Internet, faster transportation, and much better digital storage and circulation of information. A case in point is China's rapid acquisition of high-speed railway technology from Alstom, Bombardier, Siemens and the Japanese Railway Corporation. The process of acquisition of foreign railway technology started in the late 1990s and today China has more than 50 per cent of the high-speed train lines of the world. All these trains are produced in China. In addition, Chinese engineers have produced new combinations of technologies in high-speed trains.

3. In our concept, different types of organization participate in these cascades. They include large corporations and small and medium-sized enterprises (SMEs), research universities and public research organizations.

4. Knowledge flows through different channels including technology transfer, alliances and partnerships, and straight imitation. But it also can happen that several innovative organizations simultaneously arrive at similar conclusions about the feasibility of a given improved product or process. The often-mentioned example is the simultaneous invention of aluminium electrolytic production processes by the French Paul Héroult and the American Charles Hall in 1886.

5. High-tech industries often have impacts, and rely, on several sectors. Modern bioinformatics may be a great example, based on advances in both biology and informatics. Thus, innovation cascades often draw knowledge from different technologies. Cascades are not based on one technological trajectory or industry, but on several. Similarly, semiconductor, advanced materials and mechanical and robotics technologies contribute to the fast adoption of solar PV technologies.

8.1.2   Other Related Conceptions of Innovation Cascades

Bonvillian (2002) explains innovation cascades by leaps in science, such as those taking place in information and communication technology and life sciences. He rightly justifies the use of the term 'cascade', by comparison with the quiet flow of a river that best represents continuous innovation. Incremental innovation produces change but this change is slow and fairly predictable. Cascades are less so.

Delapierre and Mytelka (2003) define the innovation cascades as successive phases of innovation, the result of the strategies of monopolistic and oligopolistic firms. Those cascades put the accent on the combination of existing and previously dispersed technologies in order to arrive at a given result. They use as an example the conservation of audio and video programmes using different technologies from the video recorder to the laser videodisk.

As indicated before, we do not consider innovation cascades to be the result of large firms only; other types of innovative organizations such as SMEs, universities and government institutes can also participate in the cascade.

Sardo and Anzoise (2014) underline the ontological uncertainty that surrounds innovation cascades. The results of innovation are already fairly difficult to predict. When innovation comes in rapid sequences and many of the components of those sequences are radical innovations, then the future scenarios become unpredictable. We can integrate this hypothesis in our own thinking about cascades.

Antonelli (2008; 2009) attributes innovation cascades to the interaction between Jacobs and Marshall externalities. He argues that complementarities within a set of industries generate innovation cascades, both horizontal and vertical, and both between firms in industries and within industries. Commenting on Antonelli's contribution, Fontana (2014) argues that firms do not plan to innovate. Instead, innovation is often the result of unexpected events that change the conditions of product and factor markets as well as the amount of knowledge externalities available in the system. Also she underlines the fact that innovation may be autopoietic and self-organizing, bootstrapping and self-sustaining. This is exactly the case with innovation cascades. However, high-tech cascades do not occur just in clusters or regional innovation systems. They rather take place at all levels of the economic landscape: regional, national and international. Most innovation cascades are sectoral processes.

Innovation systems at the national, regional and sectoral level coexist (Meuer et al., 2015) and sectoral systems of innovation interact with national and regional systems (Dalitz et al., 2012). No country can expect to grow a sectoral innovation system without having previously organized a national one at least in part, and different sectoral systems interact with one another. A national incentive for renewable energy may encourage a municipal, state or provincial level to foster the creation of a cluster in their region, or adapt a national policy (that is, tax incentives) to the sub-national level of government. The rapid development of the semiconductor industry after 1960 has certainly had a positive impact on the rapid improvement of the solar photovoltaic sector, whose main component is solar cells, a specific type of semiconductor.

By applying SSI theories to explore the innovation cascade, it can be seen that innovation cascades are becoming much more frequent today for several reasons: the rise of science-based industries (Pavitt, 1984; Niosi, 2010), the increasing number of research universities in a growing number of OECD and emerging countries investing in innovation, more linkages between these loci of knowledge creation, and faster technology diffusion. Fast imitation also increases the probability of new combinations between different strands of knowledge. Innovation cascades have a definite Schumpeterian flavour.

8.1.3   The Solar Technology Innovation Cascades

As shown in previous chapters, the solar technology innovation cascade started modestly in the United States in the 1950s, with the goal of serving a niche market: the fledging satellite market. Other niche markets emerged in the following decade such as pocket calculators, and as an auxiliary power source for watches. Japanese companies were the main users and developers of solar technology for calculators and watches. Australia used and developed solar technology for remote areas where transporting electricity with cables would have been uneconomical. Then oil and gas companies added another niche market, this time for drilling in remote locations including out at sea. During several decades the technology improved, mainly that of solar cells. After 2010, there was an explosion in the numbers of new storage products. The Tesla Motors factory is one of the world's largest. It was built in partnership with Solar City, now a subsidiary of Tesla, and operated with Panasonic).[1] In 2018 the factory will produce 35 GW/hour per year of lithium batteries. The factory will create large economies of scale. There are many different types of battery but in large electrical systems, pumped storage hydro-electricity is the largest capacity form of active grid energy storage available. It represents close to 99 per cent of bulk storage capacity worldwide, close to 127 000 MW. There are other methods of storing energy, and they can be biological, chemical, electrical, electrochemical, mechanical and thermal. The US Patent and Trademark Office registered some 5725 patents on energy storage. Out of these, 237 patents also have the word 'solar' in their abstract.

There are many externalities between storage systems used for electric cars, and those systems used for storing solar energy equipment in buildings. Thus Tesla, the electric car manufacturer, has developed Tesla large home batteries, called Powerwall 2. These solar energy storage systems can provide solar energy to a house for 24 hours.

Finally, using photonics, solar glass can increase the efficiency of solar panels up to 300 per cent. One of the companies producing solar glass is HyperSolar, a US start-up. But the big producers of solar glass, both American, are Corning and Guardian. The French Saint-Gobain and British Pilkington, now a subsidiary of the Japanese NSG Group, follow them. Solar glass can magnify the efficiency of the solar panel by concentrating light into the solar cells.

8.2 GRAND CHALLENGE POLICIES

Grand challenge innovation policies are often the type of policy that fosters innovation cascades. Box 8.1 summarizes some of the most cited definitions of grand challenge policies. These are characterized by being interdisciplinary, requiring interdepartmental coordination and multi-level governance, technology convergence, cross-sectoral collaboration and long-term horizons (Cagnin et al., 2012).

8.2.1 US Policies

In 2010 the US Department of Energy launched a programme called the SunShot Initiative, a grand challenge programme whose target is reducing by 75 per cent the cost of solar power from 2010 to 2020. The Lawrence Berkeley National Laboratory (LBNL) published the evolution

---

BOX 8.1   GRAND CHALLENGE POLICIES DEFINED

'Grand Challenges are ambitious but achievable goals that harness science, technology, and innovation to solve important national or global problems and that have the potential to capture the public's imagination. Grand Challenges are an element of the President's Strategy for American Innovation.' (US Office for Science and Technology Policy, The White House, 2015)[2]

'The issues covered by the term "grand challenges" naturally lend themselves to a global outlook, are grand in scope and scale [. . .] the articulation of such grand challenges is hardly novel. The main novelty lies in the increasing attention given to such issues in formulating new missions for STI policy. The reasons for this are complex: in part, they reflect a perceived growing urgency to address a series of problems that could, if neglected, have catastrophic consequences on a global scale over the next few decades. But they also reflect a more steering of STI efforts – at least those funded by public purse – to meet explicit political goals.' (Cagnin et al., 2012)

---

of solar prices every year. The latest report published in 2016 shows that the SunShot is on track with its ambition and forecasts. The LBNL adds that the greatest gains are made in the big solar power plants. Helping the growth of solar PV installations is a 30 per cent federal investment tax credit (ITC) for solar, enacted in the Energy Policy Act of 2005; it was going to expire in 2016, but was in fact renewed. The government expects that the 27 GW of solar energy cumulatively installed up to the end of 2015 will grow to 100 GW at the end of 2020.

8.2.2 Chinese Policies

In China, policies focus on government regulation, concentrating mainly on the product diffusion and application stages, with insufficient investment in research and development in the early stage. On the other hand, however, China's PV policies are gradually changing from production supply prioritization to demand-side policy domination. According to Zhi et al. (2014), after the Renewable Energy Law was issued in 2006, policies specific to PV R&D and manufacturing enterprises began to appear, which allowed direct fiscal subsidies for technology R&D or product production of PV enterprises, tax incentives for new high-tech enterprises, R&D support from national key laboratories for enterprises, applications for financial aid for renewable energies, and so on. Since 2007, there have been policies specific to power enterprises, which drives the PV industry developed by the feed-in tariff for industrial ends and PV power generation, including additional subsidies for PV power generation feed-in tariffs and supply trading mechanisms, unified pricing, a guarantee mechanism for full-payment purchasing of renewable energy generated power and so on.

In July 2013 the China State Council issued another important policy named 'Opinions to promoting the healthy development of the photovoltaic industry', and in December 2016 the thirteenth Five-Year Plan on developing solar energy in China was issued by the National Energy Bureau. Besides the subsidies and policies to broaden demand in the domestic market, the industries are being regulated to avoid over-production and are being upgraded to higher cell-efficiency level manufacturing.

NOTES

1. https://www.pv-tech.org/editors-blog/tesla-solarcity-silveo-panasonic-1gw-buffalo-fabs-known-unknowns.
2. Available at: https://obamawhitehouse.archives.gov/administration/eop/ostp/grand-chall enges.

9. Conclusion

9.1 THEORETICAL CONTRIBUTIONS

This study has contributed four points to the academic theories and to the three levels of industrial practitioner.

9.1.1 Theoretical Contributions

By using the methodology of SSI, this study puts more flesh on the concept of the innovation cascade. Solar PV is not a classic high-tech sector, whose technology transfer pattern is quite clear in terms of location of innovation centres. The innovation cascade is described by clearly outlining the factors which show that different publication peaks appeared successively in the different regions.

The uneven development and the organizational diversity of the clusters in the same sector are proven. It was found that more diverse clusters, hosting research universities, large multi-technology corporations, public laboratories, SMEs and venture capital, such as Silicon Valley, could be more resilient than clusters based on one or two large firms. In this kind of more diverse cluster, the unusual Silicon Valley type, the exit of the anchor tenants will not lead to the decline of the cluster. The factors influencing the resilience of the clusters in this specific sector have been defined above.

The criteria for defining star scientists in the different sectors should be examined one by one together with the features of the sectors determining the contribution of academic star scientists to the development of the sector. Thus special attention should be paid to the generality of the concept and the usage of the star scientist notion.

The idea that innovation is driven by demand was explored, which may be the reason for several differences between the solar PV sector and other classic more thoroughly studied high-tech industries. This will clarify whether solar is a distinctive class of high-tech sector.

9.1.2 Contributions for Industrial Practitioners

By employing SSI methodology, the study first examined global development and innovation at country and regional levels, which can be applied to different levels of industrial practitioners.

At the country level, the evolution of the sector highlighted the competitiveness of individual countries in the past and in the present and the competitive advantages and disadvantages of different countries. This provides a solid base for policy-makers to design industrial structures and formulate viable industrial policies.

At the regional level, clusters in the world and inside major countries were identified and analysed. By understanding the differences among the clusters and the reasons for the uneven development of these clusters, policy-makers can use the experience and lessons from other clusters as well as gaining ideas on how to improve or launch their own clusters.

At the firm level, it is very important to understand the macro-environments in which the firms are positioned. The marketing classification and consumer demands were stated, the competition and innovation status were defined and the academic contributions highlighted, all of which will be inputs in the development strategies for firms either as users or directly as product or service providers.

9.2 POLICY IMPLICATIONS OF THE STUDY

As there are policy implications in the previous chapters, here we focus on the key principle to formulate the sector policies.

According to Adams et al. (2013), instead of adopting general policies with the objective of stimulating 'demand for innovation' including public procurement to regulate the solar PV sector, public policy should pay attention more to 'innovation by demand', which is to valorize the application and technological knowledge that user firms possess and to stimulate them to introduce innovations and new technologies for wider markets. It is thought that the shift in perspective from supporting demand for innovation to supporting innovation by demand could be significant, and may add an important policy input for the growth and dynamics of an economy.

The policy implications of 'innovation by demand' can be viewed in the following aspects:

● In order to push the development of the solar PV sector, firms, experts and scientists in related industries should be encouraged to use their capabilities and resources to solve the demand problems of the wider market. Only with all this integrated expertise can innovation be carried out efficiently. At the same time, large companies with related capabilities should be encouraged. For example, apart from the feed-in tariff, other supporting policies including subsidising innovation aimed at solving existing problems with solar PV technologies, funding the invention of new applications, giving priority to solar PV-related academic research, can be deployed.

- The demands of mass market electricity use should be addressed for innovation: compared to applications in extreme environments and niche markets, mass market daily electricity users have their own requirements, for example, they are more sensitive to installation and usage costs, in contrast to users in more dispersed locations, and more demanding regarding the storage of the surplus during periods of insufficient sunshine.
- The diversification of the actors in the clusters: it has been proved that clusters with diverse actors are more resilient than those with just a few agents. Along with the idea of 'innovation by demand', solar PV clusters should attract users with different applications to establish a diversified system, which will bring the sector into a healthy development cycle. As the solar PV sector is moving towards grid parity, well-understood and targeted subsidies will be critical in building the confidence of investors and attracting capital. In addition, as academic scientists are less active in technological innovation, research funds can be used to promote research and innovation in this specific domain as a priority.
- Phase out subsidies carefully. Since solar power could eventually be cost-competitive with the other conventional sources, regulators must adjust incentive structures over time, and phase them out when grid parity is reached.

9.3   LIMITATIONS AND ORIENTATION FOR FURTHER RESEARCH

Established in the theoretical framework of SSI, the study has reviewed the various elements including technologies and related innovation behaviour, firm and non-firm organizations, evolution processes and some aspects of economic performance such as geographic agglomeration. But as a comprehensive system, the sector is far from being fully understood in terms of both depth and scale.

For the aspects we have focused on during the research, we have to deepen understanding:

- In terms of the evolution of the sector, can technology transfer rules in the world be generalized for other high-tech industries? The drivers of technology transfer should be explored further.
- In terms of innovation clusters, what is the contribution of the different factors to the rise and fall of the clusters? Comparative studies with different clusters should be made.
- In terms of innovation behaviour, what are the other domains of the star scientists defined in the field? Does their academic research tend to converge to or diverge from solar PV? Can the complementary expertise from the firms help them focus more on solar PV innovation?
- In terms of entrepreneurship, as the solar tech academic entrepreneurship is limited, what features do successful entrepreneurs have? What factors influence the longevity of start-ups?
- In terms of catch-up, the review of the Chinese solar PV sector is just concentrated in its early development before 2011. Whether China can maintain the advantages at a later stage will be in the core of the real catch-up, so we need to keep track of the sector to formulate an objective answer.

For the aspects that we have not yet explored, more research on their distinctive features and their corresponding industrial performances should be carried out.

References

Abernathy, W.J. and Utterback, J.M. (1978). 'Patterns of innovation in technology', *Technology Review*, **80** (7): 1–47.

Abramovitz, M. (1986). 'Catching up, forging ahead, and falling behind', *Journal of Economic History*, **46** (2): 385–406.

Acemoglu, D. and Cao, D. (2015). 'Innovation by entrants and incumbents', *Journal of Economic Theory*, **157**: 255–94.

Achilladelis, B., Schwarzkopf, A. and Cines, M. (1990). 'The dynamics of technological innovation: the case of the chemical industry', *Research Policy*, **19** (1): 1–34.

Adams, P., Fontana, R. and Malerba, F. (2013). 'The magnitude of innovation by demand in a sectoral system: the role of industrial users in semiconductors', *Research Policy*, **42** (1): 1–14.

Agarwal, R. and Audretsch, D. (2001). 'Does entry size matter? The impact of life cycle and technology on firm survival', *Journal of Industrial Economics*, **49** (1): 21–43.

Agrawal, A. and Cockburn, I. (2003). 'The anchor tenant hypothesis: exploring the role of large, local, R&D-intensive firms in regional innovation systems', *International Journal of Industrial Organization*, **21** (9): 1227–53.

Agarwal, R., Sarkar, M.B. and Echambadi, R. (2002). 'The conditioning effect of time on firm survival: an industry life cycle approach', *Academy of Management Journal*, **45** (5): 971–94.

Agrawal, A., Cockburn, I., Galasso, A. and Oettle, A. (2014). 'Why are some regions more innovative than others? The role of small firms in the presence of large labs', *Journal of Urban Economics*, **81**: 149–55.

Ahuja, G. and Lampert, C.M. (2001). 'Entrepreneurship in the large corporation: a longitudinal study of how established firms create break-through inventions', *Strategic Management Journal*, **22**: 521–43.

Alexander, L. and Van Knippenberg, D. (2014). 'Teams in pursuit of radical innovation: a goal orientation perspective', *Academic of Management Review*, **39** (4): 423–38.

Ames, E. and Rosenberg, N. (1963). 'Changing technological leadership and industrial growth', *The Economic Journal*, **73** (289): 13–31.

Amin, A. and Robbins, K. (1990). 'The re-emergence of regional economies?

The mythical geography of flexible accumulation', *Environment and Planning D*, **8** (1): 7–34.

Amsden, A.H. and Chu, W-W. (2003). *Beyond Late Development, Taiwan's Upgrading Policies*, Boston, MA: MIT Press.

Antonelli, C. (2008). 'Pecuniary knowledge externalities: the convergence of directed technological change and the emergence of innovation systems', *Industrial and Corporate Change*, **17** (5): 1049–70.

Antonelli, C. (2009). 'Localised appropriability: pecuniary externalities in knowledge exploitation', *Technology Analysis and Strategic Management*, **21** (6): 727–42.

Archibugi, D. and Pietrobelli, C. (2003). 'The globalisation of technology and its implications for developing countries: windows of opportunity or further burden?', *Technological Forecasting and Social Change*, **70**: 861–83.

Arrow, K.J. (1962). 'Economic welfare and the allocation of resources for inventions', in R. Nelson (ed.), *The Rate and Direction of Inventive Activity: Economic and Social Factors*, Princeton, NJ: Princeton University Press, pp. 609–25.

Arthur, W.B. (1988). 'Self-reinforcing mechanisms on firm performance', in P. Anderson, K.J. Arrow and D. Pines (eds), *The Economy as an Evolving Complex System*, Redwood City, CA: Addison Wesley, pp. 9–31.

Arthur, W.B. (2009). *The Nature of Technology: What It Is and How It Evolves*, New York, NY: Free Press.

Arundel, A. and Kabla, I. (1998). 'What percentage of innovations is patented? Empirical estimates for European firms', *Research Policy*, **27** (2): 127–41.

Audretsch, D.B. (2001). 'The role of small firms in US biotechnology clusters', *Small Business Economics*, **17**: 3–15.

Audretsch, D.B. and Feldman, M. (1996). 'Innovative clusters and the industry life cycle', *Review of Industrial Organization*, **11** (2): 253–73.

Barney, J. (1991). 'Firm resources and sustained competitive advantage', *Journal of Management*, **17** (1): 99–120.

Barney, J.B. (2001). 'Resource-based theories of competitive advantage: a ten-year retrospective on the resource-based view', *Journal of Management*, **27** (6): 643–50.

Barney, J., Wright, M. and Ketchen Jr, D.J. (2001). 'The resource-based view of the firm: Ten years after 1991', *Journal of Management*, **27** (6): 625–41.

Bartel, A.P. and Lichtenberg, F.R. (1987). 'The comparative advantage of educated workers in implementing new technology', *The Review of Economics and Statistics*, **69** (1): 1–11.

Basalla, G. (1988). *The Evolution of Technology*, Cambridge: Cambridge University Press.

Bell, M. and Pavitt, K. (1992). 'Accumulating technological capability in developing countries', *The World Bank Economic Review*, **6** (supplement 1): 257–81.

Benhabib, J. and Spiegel, M.M. (1994). 'The role of human capital in economic development: evidence from aggregate cross-country data', *Journal of Monetary Economics*, **34** (2): 143–73.

Benhabib, J. and Spiegel, M.M. (2005). 'Human capital and technology diffusion', *Handbook of Economic Growth*, **1**: 935–66.

Berkers, E. and Geels, F.W. (2011). 'System innovation through stepwise reconfiguration: the case of technological transitions in Dutch greenhouse horticulture (1930–1980)', *Technology Analysis and Strategic Management*, **23** (3): 227–47.

Bessant, J., Caffyn S., Gilbert, J., Harding, R. and Webb, S. (1994). 'Rediscovering continuous improvement', *Technovation*, **14** (1): 17–29.

Block, J., Henkel, J., Schweisfurth, T.G. and Stiegler, A. (2016). 'Commercializing user innovations by vertical diversification: the user–manufacturer innovator', *Research Policy*, **45**: 244–59.

Bonabeau, E. (2002). 'Agent-based modelling: methods and techniques for simulating human systems', *Proceedings of the National Academy of Sciences of the United States of America*, **99**, suppl. 3: 7280–87.

Bonvillian, W.A. (2002). 'Science at a crossroads', *The FASB Journal*, **16**: 915–21.

Boschma, R. (2005). 'Proximity and innovation: an assessment', *Regional Studies*, **39** (1): 61–74.

Boschma, R. and Iammarino, S. (2009). 'Related variety, trade linkages and regional growth in Italy', *Economic Geography*, **85** (3): 289–311.

Boschma, R., Minondo, A. and Navarro, M. (2012). 'Related variety and regional growth in Spain', *Regional Science*, **91** (2): 241–58.

Cagnin, C., Amanatidou, E. and Keenan, M. (2012). 'Orienting European innovation systems towards grand challenges and the roles that FTA can play', *Science and Public Policy*, **39** (2): 140–52.

Cao, H. and Folan, P. (2012). 'Product life cycle: the evolution of a paradigm and literature review from 1950 to 2009', *Production Planning and Control*, **23** (8): 641–62.

Cassiman, B., Veugelers, R. and Zuniga, P. (2008). 'In search of performance effects of (in) direct industry science links', *Industrial and Corporate Change*, **17** (4): 611–46.

Castles, S. and Davidson, A. (2000). *Citizenship and Migration: Globalization and the Politics of Belonging*, New York, NY: Psychology Press.

Chaminade, C. (1999). 'Innovation processes and knowledge flows in the information and communications technologies (ICT) cluster in Spain', in OECD (ed.), *Boosting Innovation: the Cluster Approach*, Paris: OECD Publishing, pp. 219–42.

Christensen, C.M. (1997). *The Innovator's Dilemma*, Boston, MA: Harvard Business Review Press.

Christensen, C. and Bower, J.L. (1996). 'Customer power, strategic investment and the failure of leading firms', *Strategic Management Journal*, **17** (3): 197–218.

Chu, P.Y., Lin, Y.L., Hsiung, H.H. and Liu, T.Y. (2006). 'Intellectual capital: an empirical study of ITRI', *Technological Forecasting and Social Change*, **73**: 886–902.

Colalat, P.C. (2009). *Photovoltaic Systems, the Experience Curve and Learning by Doing: who is Learning and what are they Doing*, Cambridge, MA: MIT, Master of Engineering systems thesis.

Colombo, M.G. and Grilli, L. (2005). 'Founders' human capital and the growth of new technology-based firms: a competence-based view', *Research Policy*, **34** (6): 795–816.

Colombo, M.G. and Grilli, L. (2010). 'On growth drivers of high-tech start-ups: exploring the role of founders' human capital and venture capital', *Journal of Business Venturing*, **25** (6): 610–26.

Cooke, P. (2001). 'Regional innovation systems, clusters and the knowledge economy', *Industrial and Corporate Change*, **10** (4): 945–74.

Cooke, P. (2002). 'Regional innovation systems: general findings and some new evidence from biotechnology clusters', *The Journal of Technology Transfer*, **27** (1): 133–45.

Cornwall, J. (1977). *Modern Capitalism: its Growth and Transformation*, London: Martin Robertson.

Crow, M. and Bozeman, B. (1998). *Limited by Design: R&D Laboratories in the US National Innovation System*, New York, NY: Columbia University Press.

Cumming, D. and Li, D. (2013). 'Public policy, entrepreneurship, and venture capital in the United States', *Journal of Corporate Finance*, **23**: 345–67.

Dalitz, R., Holmén, M. and Scott-Kemmis, D. (2012). 'How do innovation systems interact? Schumpeterian innovation in seven Australian sectors', *Prometheus*, **30** (3): 261–89.

Dang, Y., Zhang, Y., Fan, L., Chen, H. and Roco, M.C. (2010). 'Trends in worldwide nanotechnology patent applications', *Journal of Nanoparticle Research*, **12**: 687–706.

De la Tour, A., Glachant, M. and Ménière, Y. (2011). 'Innovation and international technology transfer: the case of the Chinese photovoltaic industry', *Energy Policy*, **39**: 761–70.

Delapierre, M. and Mytelka, L. (2003). 'Cascades d'innovations et nouvelles stratégies oligopolistiques', *Revue d'économie Industrielle*, **103**: 233–52.

Dosi, G. (1982). 'Technological paradigms and technological trajectories: a suggested interpretation of the determinants of technical change', *Research Policy*, **11** (3): 147–62.

Dosi, G. and Soete, L. (1988). 'Technical change and international trade', in G. Dosi, C. Freeman, R. Nelson, G. Silverberg and L. Soete (eds), *Technical Change and Economic Theory*, London: Pinter Publishers.

Dosi, G., Pavitt, K. and Soete, L. (1990). *The Economics of Technical Change and International Trade*, Pisa: Sant'Anna School of Advanced Studies.

European Photovoltaic Industry Association (2014). 'Global market outlook for photovoltaics 2014–2018', available at: http://www.epia.org/fileadmin/user_upload. Publications/44_epia_gmo_report_ver_ 17_mr.pdf.

Eurostat (2004). *Innovation in Europe: Results for the EU, Iceland and Norway*, Luxembourg: European Communities.

Farmer, J.D. and Foley, D. (2009). 'The economy needs agent-based modelling', *Nature*, **460** (7256): 685–6.

Feldman, M. (2003). 'The locational dynamic of the US biotech industry: knowledge externalities and the anchor hypothesis', *Industry and Innovation*, **10** (3): 311–28.

Flamm, K. (1988). *Creating the Computer: Government, Industry and High Technology*, Washington, DC: The Brookings Institution.

Fontana, M. (2014). 'Pluralism (s) in economics: lessons from complexity and innovation. A review paper', *Journal of Evolutionary Economics*, **24** (1): 189–204.

Foster, A.D. and Rosenzweig, M.R. (1995). 'Learning by doing and learning from others: human capital and technical change in agriculture', *Journal of Political Economy*, **103** (6): 1176–209.

Freeman, C. (2002). 'Continental, national and sub-national innovation systems: complementarity and economic growth', *Research Policy*, **31** (2): 191–211.

Frenken, K., Cefis, E. and Stamm, E. (2015). 'Industrial dynamics and clusters: a survey', *Regional Studies*, **49** (1): 10–27.

Frenken, K., Oort, V.F. and Verburg, T. (2007). 'Related variety, unrelated variety and regional economic growth', *Regional Studies*, **41**: 685–97.

Fu, X. (2015). *China's Path to Innovation*, Cambridge: Cambridge University Press.

Fuller, A.W. and Rothaermel, F.T. (2012). 'When stars shine: the effects of faculty founders on new technology ventures', *Strategic Entrepreneurship Journal*, **6**: 220–35.

Garavaglia, C., Malerba, F., Orsenigo, L. and Pezzoni, M. (2012). 'Technological regimes and demand structure in the evolution of the pharmaceutical industry', *Journal of Evolutionary Economics*, **22** (4): 677–709.

Gault, F. and Von Hippel, E.A. (2009). 'The prevalence of user innovation and free innovation transfers: implications for statistical indicators and innovation policy', MIT Sloan School Working Paper No. 4722-09.

Geels, F.W. (2002). 'Technological transitions as evolutionary reconfiguration processes: a multi-level perspective and a case study', *Research Policy*, **31** (8–9): 1257–74.

Geels, F.W. (2004). 'From sectoral systems of innovation to socio-technical systems. Insights about dynamics and change from sociology and institutional theory', *Research Policy*, **33**: 897–920.

Geroski, P.A. (2003). *The Early Evolution of Markets*, New York, NY: Oxford University Press.

Gerschenkron, A. (1962). *Economic Development in Historical Perspective*, Cambridge, MA: Belknap Press of Harvard University.

Gibbons, M. and Littler, D. (1979). 'The development of an innovation: the case of Porvair', *Research Policy*, **8** (1): 2–25.

Gittelman, M. (2008). 'A note on the value of patents as indicators of innovation: implications for management research', *The Academy of Management Perspectives*, **22** (3): 21–7.

Global Trends in Renewable Energy Investment (2015). Frankfurt School-UNEP Centre, Frankfurt am Main, available at: http://www.fs-unep-centre.org.

Glover, P.C. (2013). 'Europe renewable hype implodes as Germany solar goes belly up', *Energy Tribune*, 24 June, p. 5.

Gomulka, S. (1971). *Inventive Activity, Diffusion, and the Stages of Economic Growth* (Vol. 24), Aarhus: Aarhus University, Institute of Economics.

Gould, S.J. and Eldredge, N. (1977). 'Punctuated equilibria: the tempo and mode of evolution reconsidered', *Paleobiology*, **3** (2): 115–51.

Granstrand, O., Patel, P. and Pavitt, K. (1997). 'Multi-technology corporations: why they have "distributed" rather than "distinctive core" competencies', *California Management Review*, **39** (4): 8–27.

Grantham, M.L. (1997). 'The validity of the product life cycle in the high-tech industry', *Marketing Intelligence & Planning*, **15** (1): 4–10.

Grau, T., Hu, M. and Neuhaff, K. (2012). 'Survey of photovoltaic industry and policy in Germany and China', *Energy Policy*, **51** (12): 20–37.

Green, M. (2009). 'The path to 25% silicon solar cell efficiency: history of silicon cell evolution', *Progress in Photovoltaics: Research and Applications*, **17**: 183–9.

Green, M.A. (2015). 'Corrigendum to "Solar cell efficiency tables (version 46) in *Progress in Photovoltaics: Research and Applications*, **23** (7): 805–12"', *Progress in Photovoltaics: Research and Applications*, **23** (9): 1202.

Green, S.G., Gavin, M.B. and Aiman-Smith, L. (1995). 'Assessing the multidimensional measure of radical technological innovation', *IEEE Transactions in Engineering Management*, **42** (3): 203–14.

Greene, M. (2007). 'The demise of the lone author', *Nature*, **450** (7173): 1165.

Grossman, G.M. and Helpman, E. (1991). 'Quality ladders in the theory of growth', *The Review of Economic Studies*, **58** (1): 43–61.

Groysberg, B., Lee, L.E. and Nanda, A. (2008). 'Can they take it with them? The portability of star knowledge workers' performance', *Management Science*, **54** (7): 1213–30.

Habakkuk, H.J. (1962). 'American and British technology in the 19th century', *Economica, New Series*, **118** (30): 215–17.

Hall, B.H., Jaffe, A.B. and Trajtenberg, M. (2000). 'Market value and patent citations: a first look', No. W 7741, Cambridge, MA: National Bureau of Economic Research.

Hamel, G. and Heene, A. (1994). *Competence-based Competition*, Chichester: John Wiley and Sons.

Han, X. and Niosi, J. (2016). 'Star scientists in PV technology and the limits of academic entrepreneurship', *Journal of Business Research*, **69** (5): 1707–11.

Hassink, R. (2010). 'Regional resilience: a promising concept to explain differences in regional economic adaptability?', *Cambridge Journal of Regions, Economy and Society*, **3** (1): 45–58.

Haupt, R., Kloyer, M. and Lange, M. (2007). 'Patent indicators for the technology life cycle development', *Research Policy*, **36** (3): 387–98.

Hayter, C.S. (2015). 'Social networks and the success of university spin-offs toward an agenda for regional growth', *Economic Development Quarterly*, **29** (1): 3–13.

Hienerth, C. (2006). 'The commercialization of user innovations: the development of the rodeo kayak industry', *R&D Management*, **36** (3): 273–94.

Hitt, M.A., Biermant, L., Shimizu, K. and Kochhar, R. (2001). 'Direct and moderating effects of human capital on strategy and performance in professional service firms: a resource-based perspective', *Academy of Management Journal*, **44** (1): 13–28.

Hobday, M. (1995). 'East Asian latecomer firms: learning the technology of electronics', *World Development*, **23** (7): 1171–93.

Hoser, N. (2013). 'Public funding in the academic field of nanotechnology:

a multi-agent-based model', *Computational and Mathematical Organization Theory*, **19**: 253–81.

Hsu, P.H., Shyu, J.Z., Yu, H.C., You, C.C. and Lo, T.H. (2003). 'Exploring the interaction between industrial incubators and industrial clusters: the case of the ITRI incubator in Taiwan', *R&D Management*, **33** (1): 79–91.

Hu, A.G. and Jaffe, A.B. (2003). 'Patent citations and international knowledge flow: the cases of Korea and Taiwan', *International Journal of Industrial Organization*, **21** (6): 849–80.

Hu, M.C. (2008). 'Knowledge flows and innovation capability: the patenting trajectory of Taiwan's thin film transistor-liquid crystal display industry', *Technological Forecasting and Social Change*, **75** (9): 1423–38.

International Energy Agency (IEA) (2014). 'How solar energy could be the largest source of electricity by mid-century'. Press Release, 29 September, available at: http://www.iea.org/newsroomandevents/pressreleases/2014/september/name-125873-en.html.

Jacobsson, S., Sandén, B. and Bangens, L. (2004). 'Transforming the energy system: the evolution of the German technological system for photovoltaics', *Technology Analysis and Strategic Management*, **16** (1): 3–30.

Jaffe, A.B. and Trajtenberg, M. (2002). *Patents, Citations & Innovations: A Window on the Knowledge Economy*, Cambridge, MA: MIT Press.

Jang, S.L., Lo, S. and Chang, W.H. (2009). 'How do latecomers catch up with forerunners? Analysis of patents and patent citations in the field of flat panel display technologies', *Scientometrics*, **79** (3): 563–91.

Järvenpää, H.M., Mäkinen, S.J. and Seppänen, M. (2011). 'Patent and publishing activity sequence over a technology's life cycle', *Technological Forecasting and Social Change*, **78** (2): 283–93.

Kenney, M. (2000). *Understanding Silicon Valley: the Anatomy of an Entrepreneurial Region*, Stanford, CA: Stanford University Press.

Ketelhöhn, N.W. (2006). 'The role of clusters as sources of dynamic externalities in the US semiconductor industry', *Journal of Economic Geography*, **6** (5): 679–99.

Kim, L. (1980). 'Stages of development of industrial technology in a developing country: a model', *Research Policy*, **9** (3): 254–77.

Kim, L. (1997a). *Imitation to Innovation: The Dynamics of Korea's Technological Learning*, Boston, MA: Harvard Business School Press.

Kim, L. (1997b). 'The dynamics of Samsung's technological learning in semiconductors', *California Management Review*, **39** (3): 86–100.

Kim, L. (1998). 'Technology policies and strategies for developing countries: lessons from the Korean experience', *Technology Analysis and Strategic Management*, **10** (3): 311–24.

Kim, L. and Nelson, R.R. (2000). *Technology, Learning, and Innovation:*

Experiences of Newly Industrializing Economies, Cambridge, MA: Cambridge University Press.

Kirkegaard, J.F., Hanemann, T., Weischer, L. and Miller, M. (2010). 'Toward a sunny future? Global integration in the solar PV industry', Peterson Institute for International Economics Working Paper, May.

Kitschelt, H. (1991). 'Industrial governance structures, innovation strategies and the case of Japan: sectoral or cross-national comparative analysis?', *International Organization*, **45** (4): 453–93.

Klepper, S. (1997). 'Industry life cycles', *Industrial and Corporate Change*, **6** (1): 145–81.

Klepper, S. and Thompson, P. (2007). 'Spin-off entry into high-tech industries: motives and consequences', in F. Malerba and S. Brusoni (eds), *Perspectives on Innovation*, Cambridge, MA: Cambridge University Press, pp. 187–218.

Korhonen, J. (2001). 'Coproduction of heat and power: an anchor tenant of a regional industrial eco-system', *Journal of Cleaner Production*, **9**: 509–17.

Korhonen, J. and Snakin, J.P. (2001). 'An anchor tenant approach to network management: considering regional material and energy flows models', *International Journal of Environmental Technology and Management*, **1** (4): 444–63.

Kortum, S. and Lerner, J. (1999). 'What is behind the recent surge in patenting', *Research Policy*, **28**: 1–22.

Kumar, N., Scheer, L. and Koller, P. (2000). 'From market driven to market driving', *European Management Review*, **18** (2): 129–42.

Kupriyanov, V.S., Anilova, N.Z. and Kachyrova, V. (2014). 'Nanotechnology as a factor in the development of regional clusters', *Journal of Applied Engineering Science*, **12** (4): 285–90.

Lambkin, M. and Day, G.S. (1989). 'Evolutionary processes in competitive markets: beyond the product life cycle', *Journal of Marketing*, **53** (3): 4–20.

Landström, H., Harirchi, G. and Åstrom, F. (2012); 'Entrepreneurship: exploring the knowledge base', *Research Policy*, **41** (7): 1154–81.

Lane, D. (2003). *Towards an Agenda for Social Innovation*, Venice: European Centre for Living Technology.

Lane, D. (ed.) (2009a). *Complexity Perspectives in Innovation and Social Change*, Dordrecht: Springer.

Lane, D. (2009b). 'Complexity and economic dynamics', in C. Antonelli (ed.), *Handbook on the Economic Complexity of Technological Change*, Cheltenham, UK and Northampton, MA, USA: Edward Elgar Publishing, pp. 63–80.

Lane, D. (2012). 'The dynamics of innovation cascades', paper presented at the biannual congress of the International Schumpeter Society, Brisbane, Australia.

Lane, D. (2016). 'Innovation cascades: artefacts, organization and attributions', *Philosophical Transactions of the Royal Society B*, **371** (1690): 2015.0194.

Lane, D. and Maxfield, R. (1996). 'Strategy under complexity: fostering regenerative relationships', *Long Range Planning*, **29** (2): 215–31.

Larsen, P.O. and Von Ins, M. (2010). 'The rate of growth in scientific publication and the decline of coverage provided by SCI', *Scientometrics*, **84**: 575–603.

Lee, J. and Veloso, F.M. (2008). 'Interfirm innovation under uncertainty: empirical evidence for strategic knowledge partitioning', *Journal of Product Innovation Management*, **25** (5): 418–35.

Lee, K. (2005). 'Making a technological catch-up: barriers and opportunities', paper presented at the 2005 Asialics Conference, Jeju, Korea, April.

Lee, K. and Jin, J. (2012). 'From learning knowledge outside to creating knowledge within: Korea's mobile phone industry compared with those of Japan, Taiwan and China', in J. Machlich and W. Pascha (eds), *Korean Science and Technology in an International Perspective*, Berlin: Physica-Verlag HD, pp. 197–218.

Lee, K. and Kim, M. (2004). *The Rise of China and the Korean Firms Seeking New Divisions of Labor*, paper presented at the international conference on China and East Asia, organized by the KIEP.

Lee, K. and Lim, C. (2001). 'Technological regimes, catching-up and leapfrogging: findings from the Korean industries', *Research Policy*, **30**: 459–83.

Lee, K. and Yoon, M. (2010). 'International, intra-national and inter-firm knowledge diffusion and technological catch-up: the USA, Japan, Korea and Taiwan in the memory chip industry', *Technology Analysis and Strategic Management*, **22** (5): 553–70.

Lee, K. and Wang, R. (2010). 'Science and technology institutions and performance in China: the semi-conductor industry', in X. Huang (ed.), *The Institutional Dynamics of China's Great Transformation*, London: Routledge, pp. 55–77.

Liao, S.W. and Xu, H.J. (2012). 'Comparison of the governmental policies on the PV industry, regional and industrial economics', *China Business, Economic Theoretical Research*, **3**: 15–17.

Link, A.N., Scott, J.T. and Siegel, D.S. (2003). 'The economics of intellectual property at universities: an overview of the special issue', *International Journal of Industrial Organization*, **21** (9): 1217–25.

Liu, X. (2008). 'China's development model: an alternative strategy for technological catch-up', SLPTMD Working Paper Series No. 020, Oxford University.

Low, M.B. and Abrahamson, E. (1997). 'Movements, bandwagons and clones: industry evolution and the entrepreneurial process', *Journal of Business Venturing*, **12** (5): 435–57.

Lowe, R.A. and Gonzalez-Brambila, C. (2007). 'Faculty entrepreneurs and research productivity', *Journal of Technology Transfer*, **32**: 173–94.

Lundvall, B.A. (1992). *National Systems of Innovation.* London: Pinter.

Luo, S., Lovely, M. and Popp, D. (2014). 'Intellectual returnees as drivers of indigenous innovation: evidence from the Chinese photovoltaic industry', NBER Working Paper No. 19158, Cambridge, MA.

Ma, L.L. (2012). 'Development trends of solar photovoltaic industry', report submitted to the Chinese Consultancy Board of GEM, Beijing.

Mahalingam, A. and Reiner, D. (2016). 'Energy subsidies at times of economic crisis: a comparative study and scenario analysis of Italy and Spain', Energy Policy Research Group Working Paper No. 1063 and Cambridge Working Paper in Economics No. 1068.

Malerba, F. (2002). 'Sectoral systems of innovation and production', *Research Policy*, **31** (2): 247–64.

Malerba, F. (2004). *Sectoral Systems of Innovation*, Cambridge: Cambridge University Press.

Malerba, F. (2006). 'Innovation and the evolution of industries', *Journal of Evolutionary Economics*, **16**: 3–23.

Malerba, F. (2007). 'Innovation and the dynamics and evolution of industries: progress and challenges', *International Journal of Industrial Organization*, **25** (4): 675–99.

Malerba, F. and Orsenigo, L. (1996). 'The dynamics and evolution of industries', *Industrial and Corporate Change*, **5** (1): 51–88.

Malerba, F. and Orsenigo, L. (2002). 'Innovation and market structure in the dynamics of the pharmaceutical industry and biotechnology: towards a history-friendly model', *Industrial and Corporate Change*, **11** (4): 667–703.

Malerba, F. and Orsenigo, L. (2015). 'The evolution of the pharmaceutical industry', *Business History*, **57** (5): 664–87.

Malerba, F., Nelson, R., Orsenigo, L. and Winter, S. (1999). 'History-friendly models of industrial evolution: the computer industry', *Industrial and Corporate Change*, **1**: 3–41.

Malerba, F., Nelson, R., Orsenigo, L. and Winter, S. (2001). 'Competition and industrial policy in a history-friendly model of the evolution of the computer industry', *International Journal of Industrial Organization*, **19**: 635–64.

Mangematin, V. and Errabi, K. (2012). 'The determinants of science-based cluster growth: the case of nanotechnology', *Environment and Planning C*, **30**: 128–46.

Mansfield, E. (1986). 'Patents and innovation: an empirical study', *Management Science*, **32** (2): 173–81.

Martin, R. and Sunley, P. (2003). 'Deconstructing clusters: chaotic concept or policy panacea', *Journal of Economic Geography*, **3**: 5–35.

Martin, R. and Sunley, P. (2015). 'On the notion of regional economic resilience: conceptualization and explanation', *Journal of Economic Geography*, **15** (1): 1–42.

Maskell, P. and Malmberg, A. (2007). 'Myopia, knowledge development and cluster evolution', *Journal of Economic Geography*, **7** (5): 603–18.

Mathews, J.A. (2006). 'Catch-up strategies and the latecomer effect in industrial development', *New Political Economy*, **11** (3): 313–35.

Mathews, J.A., Hu, M.C. and Wu, C.Y. (2011). 'Fast-follower industrial dynamics: the case of Taiwan's emergent solar photovoltaic industry', *Industry and Innovation*, **18** (02): 177–202.

Mazzoleni, R. and Nelson, R.R. (2007). 'Public research institutions and economic catch-up', *Research Policy*, **36** (10): 1512–28.

McGahan, A.M. and Silverman, B.S. (2001). 'How does innovative activity change as industries mature?', *International Journal of Industrial Organization*, **19** (7): 1141–60.

McIntyre, S.H. (1988). 'Market adaptation as a process in the product life cycle of radical innovations and high technology products', *Journal of Product Innovation Management*, **5** (2): 140–49.

McKelvey, M.D. (1996). *Evolutionary Innovations*, Oxford: Oxford University Press.

McKelvey, M. (1998). 'Evolutionary innovations: learning, entrepreneurship and the dynamics of the firm', *Journal of Evolutionary Economics*, **8** (2): 157–75.

McKelvey, M. and Niosi, J. (2015). 'Innovation cascades and the emergence of the bio-economy', in *2015 Portland International Conference on Management of Engineering and Technology (PICMET)*, 2–6 August, Portland, OR: IEEE, pp. 545–50.

McMillan, G.S., Narin, F. and Deeds, D.L. (2000). 'An analysis of the critical role of public science in innovation: the case of biotechnology', *Research Policy*, **29** (1): 1–8.

Menzel, M.P. and Fornahl, D. (2009). 'Cluster life cycles: dimensions and rationales of cluster evolution', *Industrial and Corporate Change*, **6**: 1–34.

Mercom Capital Group (2015). *Global Solar investments*, Austin, TX.

Metcalfe, J.S., Foster, J. and Ramiogan, R. (2006). 'Adaptive economic growth', *Cambridge Journal of Economics*, **30** (1): 7–32.

Meuer, J., Rupietta, C. and Backes-Gellner, U. (2015). 'Layers of co-existing innovation systems', *Research Policy*, **44** (4): 888–910.

Mokyr, J. (2002). *The Gift of Athena, Historical Origins of the Knowledge Economy*, Princeton, NJ: Princeton University Press.

Moretti, E. and Wilson, D.J. (2013). 'State incentives for innovation, star scientists and jobs: evidence from biotech', Cambridge, MA: NBER Working Paper Series No.19294.

Mowery, D. (ed.) (1996). *The International Computer Software Industry. A Comparative Study of Industry Evolution and Structure*, Oxford, UK and New York, NY: Oxford University Press.

Mowery, D.C., Nelson, R.R., Sampat, B.N. and Ziedonis, A.A. (2001). 'The growth of patenting and licensing by US universities: an assessment of the effects of the Bayh–Dole Act of 1980', *Research Policy*, **30**: 99–119.

Nelson, R.R. (2005). *Technology, Institutions and Economic Growth*, Cambridge, MA: Harvard University Press.

Nelson, R.R. (2006). 'Evolutionary social science and universal Darwinism', *Journal of Evolutionary Economics*, **16**: 491–510.

Nelson, R.R. and Phelps, E.S. (1966). 'Investment in humans, technological diffusion, and economic growth', *The American Economic Review*, **56** (1/2): 69–75.

Nelson, R.R. and Winter, S.G. (1982a). 'The Schumpeterian trade-off revisited', *The American Economic Review*, **72** (1): 114–32.

Nelson, R.R. and Winter, S.G. (1982b). *An Evolutionary Theory of Economic Change*, Cambridge, MA: Belknap Press of Harvard University.

Niosi, J. (2000). 'Science-based industries: a new Schumpeterian taxonomy', *Technology in Society*, **22**: 429–44.

Niosi, J. (2005). *Canada's Regional Innovation Systems*, Montreal and Kingston: McGill-Queen's University Press.

Niosi, J. (2010). *Building National and Regional Innovation Systems: Institutions for Economic Development*, Cheltenham, UK and Northampton, MA, USA: Edward Elgar Publishing.

Niosi, J. and Banik, M. (2005). 'The evolution and performance of biotechnology regional systems of innovation', *Cambridge Journal of Economics*, **29**: 343–57.

Niosi, J. and Bas, T.G. (2001). 'The competencies of regions: Canada's clusters in biotechnology', *Small Business Economics*, **17**: 31–42.

Niosi, J. and Queenton, J. (2010). 'Knowledge capital in biotechnology industry: impacts on Canadian firm performance', *International Journal of Knowledge-Based Development*, **1** (1–2): 136–51.

Niosi, J. and Reid, S.E. (2007). 'Biotechnology and nanotechnology: science-based industries as windows of opportunity for LDCs?', *World Development*, **35** (3): 426–38.

Niosi, J. and Zhegu, M. (2010). 'Anchor tenants and regional innovation systems: the aircraft industry', *International Journal of Technology Management*, **50** (3/4): 263–85.

Niosi, J., Saviotti, P., Bellon, B. and Crow, M. (1993). 'National systems of innovation: in search of a workable concept', *Technology in Society*, **15** (2): 207–27.

O'Connor, G.C. and McDermott, C.M. (2004). 'The human side of radical innovation', *Journal of Engineering and Technology Management*, **21**: 11–30.

O'Connor, G.C. and Rice, M.P. (2013). 'A comprehensive model of uncertainty associated with radical innovation', *Journal of Product Innovation Management*, **30** (Suppl. 1): 2–18.

OECD (1992). *Technology and the Economy*, Paris: OECD Publishing.

OECD (2001). *Innovative Clusters: Drivers of National Innovation Systems*, Paris: OECD Publishing.

Oettl, A. (2012). 'Reconceptualizing stars: scientist helpfulness and peer performance', *Management Science*, **58** (6): 1122–40.

Østergaard, C.R. and Park, E. (2015). 'What makes cluster decline? A study on disruption and evolution of a high-tech cluster in Denmark', *Regional Studies*, **49** (5): 834–49.

Pack, H. and Westphal, L.E. (1986). 'Industrial strategy and technological change: theory versus reality', *Journal of Development Economics*, **22** (1): 87–128.

Parida, B., Iniyan, S. and Goic, R. (2011). 'A review of solar photovoltaic technologies', *Renewable and Sustainable Energy Reviews*, **15** (3): 1625–36.

Pavitt, K. (1984). 'Sectoral patterns of technical change: towards a taxonomy and a theory', *Research Policy*, **13** (6): 343–74.

Penrose, E. (1959). *The Theory of the Firm*, New York, NY: John Wiley and Sons.

Perez, C. (1983). 'Structural change and the assimilation of new technologies in the economic and social system', *Future*, **15** (5): 357–75.

Perez, C. (2011). *Techno-economic Paradigms: Essays in Honour of Carlota Perez*, London: Anthem Press.

Perez, C. and Soete, L. (1988). 'Catching up in technology: entry barriers and windows', in G. Dosi, R. Nelson, G. Silverberg and L. Soete (eds), *Technical Change and Economic Theory*, London: Pinter, pp. 458–79.

Perlin, J. (2002). 'The story of solar concentrators', in *Proceedings of the First International Conference on Solar Electric Concentrators: A Joint Conference with the 29th IEEE PVSC*.

Perroux, F. (1972). 'Le pouvoir a droit de cité en science économique', *Le Monde*, 26 September, p. 10.

Perroux, F. (1982). *Dialogue des Monopoles et des Nations*, Grenoble: Presses de l'Université de Grenoble.

Petroski, H. (1994). *The Evolution of Useful Things*, New York, NY: Random House.

Pew Charitable Trust (2014). *Who's Winning the Clean Energy Race?* Accessed on 13 May 2014 at: http://www.pewenvironment.org/upload edFiles/PEG/Publications/Report/clen-whos-winning-the-clean-energy-race-2013.pdf.

Pirnay, F., Surlemont, B. and Nlemvo, F. (2000). 'Towards a typology of university spin-offs', *Small Business Economics*, **21**: 355–69.

Porter, M.E. (1990). 'The competitive advantage of nations', *Harvard Business Review*, **68** (2): 73–93.

Porter, M.E. (1998). 'Cluster and the new economics of competition', *Harvard Business Review*, November–December.

Porter, M.E. (2000). 'Location, competition and economic development: local clusters in a global economy', *Economic Development Quarterly*, **14** (1): 15–34.

Porter, M.E. (2003). 'The economic performance of regions', *Regional Studies*, **37**: 549–78.

Porter, M.E. and Stern, S. (2000). 'Measuring the "ideas" production function: evidence from international patent output', NBER Working Paper No. 7891, Cambridge, MA: National Bureau of Economic Research.

Posner, M.V. (1961). 'International trade and technical change', *Oxford Economic Papers*, **13** (3): 323–41.

Powers, J.B. and McDougall, P. (2005). 'University start-up formation and technology licensing with firms that go public: a resource-based view of academic entrepreneurship', *Journal of Business Venturing*, **20**: 291–311.

Ritala, P. and Hurmelinna Laukkanen, P. (2013). 'Incremental and radical innovation in coopetition: the role of absorptive capacity and appropriability', *Journal of Product Innovation Management*, **30** (1): 154–69.

Robertson, P.L. and Langlois, R.N. (1995). 'Innovation, networks and vertical integration', *Research Policy*, **24** (4): 543–62.

Roco, M.C. and Bainbridge, W.S. (2005). 'Societal implications of nanoscience and nanotechnology: maximizing human benefit', *Journal of Nanoparticle Research*, **7**: 1–13.

Rothwell, R. (1980). 'The impact of regulation on innovation', *Technological Forecasting and Social Change*, **17** (1): 7–34.

Rothwell, R. and Wissema, H. (1986). 'Technology, culture and public policy', *Technovation*, **4** (2): 91–115.

Sahu, B.K. (2015). 'A study on global solar PV energy developments and policies with special focus on the top ten solar PV power producing countries', *Renewable and Sustainable Energy Reviews*, **43** (6): 621–34.

Sainio, L.M., Ritala, P. and Hurmelinna-Laukkanen, P. (2012). 'Constituents of radical innovation: exploring the role of strategic orientations and market uncertainty', *Technovation*, **32**: 591–9.

Sapsalis, E., Van Pottelsberghe, B. and Navon, R. (2006). 'Academic and industry patenting: an in-depth analysis of what determines patent value', *Research Policy*, **35**: 1631–45.

Sardo, S. and Anzoise, V. (2014). 'Managing organization uncertainties: dynamic evaluation and processes design', paper presented at the conference *A Matter of Design. Making Society through Science and Technology*, June, Milan.

Saviotti, P.P. (1996). *Technological Evolution, Variety and the Economy*, Cheltenham, UK and Brookfield, VT, USA: Edward Elgar Publishing.

Saviotti, P.P. and Pyka, A. (2004). 'Economic growth by the creation of new sectors', *Journal of Evolutionary Economics*, **14**: 1–35.

Saxenian, A. (1994). *Regional Advantage: Culture and Competition in Silicon Valley and Route 128*, Boston, MA: Harvard University Press.

Saxenian, A. and Hsu, J.Y. (2001). 'The Silicon Valley–Hsinchu connection: technical communities and industrial upgrading', *Industrial and Corporate Change*, **10** (4): 893–920.

Schultz, P. (2011). 'Conservation means behaviour', *Conservation Biology*, **25** (6): 1080–83.

Schumpeter, J. (1916). *The Theory of Economic Development*, Boston, MA: Harvard University Press.

Schumpeter, J. (1939). *Business Cycles*, New York, NY, USA, Toronto, Canada and London, UK: McGraw-Hill.

Scoville, W.C. (1951). 'Minority migrations and the diffusion of technology', *The Journal of Economic History*, **11** (4): 347–60.

Shah, S.K. and Tripsas, M. (2007). 'The accidental entrepreneur: the emergent and collective process of user entrepreneurship', *Strategic Entrepreneurship Journal*, **1** (1–2): 123–40.

Shapira, P. and Youtie, J. (2008). 'Emergency of nano districts in the United States: path dependency or new opportunities?', *Economic Development Quarterly*, **22** (3): 187–99.

Shyu, J.Z. and Chiu, Y.C. (2002). 'Innovation policy for developing Taiwan's competitive advantages', *R&D Management*, **32** (4): 369–74.

Singh, G.K. (2013). 'Solar power generation by PV (photovoltaic) technology: a review', *Energy*, **53**: 1–13.

Sorensen, J.B. (2007). 'Bureaucracy and entrepreneurship: workplace effects on entrepreneurial entry', *Administrative Science Quarterly*, **52** (3): 387–412.

Souder, W.E. (1983). 'Organizing for modern technology and innovation: a review and synthesis', *Technovation*, **2** (1): 27–44.

Statistics Canada (2014). 'Survey of intellectual property commercialization in the higher education sector', Ottawa.

Sternberg, R. (2014). 'Success factors of university spin-offs: regional government support versus regional environment', *Technovation*, **34** (3): 137–48.

Suenaga, K. (2015). 'The emergence of technological paradigms: the evolutionary process of science and technology in economic development', in A. Pyka and J. Foster (eds), *The Evolution of Economic and Innovation Systems*, Heidelberg, Germany, New York, NY, USA, Dordrecht, the Netherlands and London, UK: Springer, pp. 211–28.

Suire, R. and Vicente, J. (2009). 'Why do some places succeed when others decline? A social interaction model of cluster viability', *Journal of Economic Geography*, **9**: 381–404.

Svetiev, Y. (2011). 'The role of intellectual property in joint innovation and development', Working Paper Series 2011/36, Florence: European University Institute.

Swann, G.M.P., Prevezer, M. and Stout, D. (1998). *The Dynamics of Industrial Clustering: International Comparisons in Computing and Biotechnology*, Oxford: Oxford University Press.

Tartari, V., Perkmann, M. and Salter, A. (2014). 'In good company: the influence of peers on industry engagement by academic scientists', *Research Policy*, **43** (7): 1189–203.

Tassey, G. (1991). 'The functions of technology infrastructure in a competitive economy', *Research Policy*, **20** (4): 345–61.

Teubal, M. (1997). 'A catalytic and evolutionary approach to horizontal technology policies', *Research Policy*, **25**: 1161–88.

Trippl, M. (2011). 'Scientific mobility and knowledge transfer at the interregional and intraregional level', *Regional Studies*, **47** (10): 1653–67.

Tushman, M.L. and Romanelli, E. (2008). 'Organizational evolution', in W.W. Burke, D.G. Lake and J.W. Paine (eds), *Organizational Change*, San Francisco, CA: Jossey-Bass.

Tzabbar, D. and Kehoe, R.R. (2014). 'Can opportunity emerge from disarray? An examination of exploration and exploitation following star scientist turnover', *Journal of Management*, **40** (2): 449–82.

US Department of Energy (2010). 'The history of solar', available at: https://www1.eere.energy.gov/solar/ pdfs/solar_timeline.pdf.

US National Academy of Science Panel on Electricity from Renewable Resources (2010). *Electricity from Renewable Resources: Status Prospects and Impediments*, Washington, DC: National Academies Press.

US Office of Science and Technology, The White House (2015), '21st century grand challenges', available at: https://obamawhitehouse.archives.gov/administration/eop/ostp/grand-challenges.

Utterback, J. (1994). *Mastering the Dynamics of Innovation*, Boston, MA: Harvard Business School Press.

Vega, A. (2013). 'Siemens closes down last solar energy business after failing to sell', *Energy and Technology Magazine*, 18 June, available at: http://eandt.theiet.org/news/2013/jun/siemens-closing-solar.cfm.

Vidican, G., Woon, W.L. and Madnick, S. (2009). 'Measuring innovation using bibliometric techniques: the case of solar photovoltaic industry', *MIT Sloan School Working Paper No. 4733-09*.

Von Hippel, E. (1986). 'Lead users: a source of novel product concepts', *Management Science*, **32** (7): 791–805.

Von Hippel, E. (2005). *Democratizing Innovation: The Evolving Phenomenon of User Innovation*, Boston, MA: The MIT Press.

Von Tunzelmann, G.N. (1978). *Steam Power and British Industrialization to 1860*, Oxford: Oxford University Press.

Wagner, C. and Leydesdorff, L. (2005). 'Network structure, self-organization and the growth of international collaboration in science', *Research Policy*, **34** (10): 1608–18.

Walter, A., Auer, M. and Ritter, T. (2006). 'The impact of network capabilities and entrepreneurial orientation on university spin-off performance', *Journal of Business Venturing*, **21**: 541–67.

Welch, F. (1975). 'Human capital theory: education, discrimination, and life cycles', *The American Economic Review*, **65** (2): 63–73.

Whitley, R. (2003). 'Competition and pluralism in the public sciences: the impact of institutional frameworks on the organization of academic science', *Research Policy*, **32** (6): 1015–29.

Winder, N. (2007). 'Innovation and metastability: a system model', *Ecology and Society*, **12** (2): 28.

Wright, M., Birley, S. and Mosey, S. (2004). 'Entrepreneurship and university technology transfer', *Journal of Technology Transfer*, **29** (3–4): 235–46.

Wu, C.Y. and Mathews, J. (2012). 'Knowledge flows in the solar photovoltaic industry: insights from patenting by Taiwan, Korea, and China', *Research Policy*, **41**: 524–40.

Yakushiji, T. (1986). 'Technological emulation and industrial development', paper presented at the conference on Innovation diffusion, 17–21 March, Venice.

Yin, Z.Q. and solar energy work team of Chinese Academy of Engineering (2011). *Renewable Energy Volume of Middle and Long-term (2030, 2050). Development Strategy of Chinese Energy Industry Serials*. Chinese Science Press.

Youtie, J. and Shapira, P. (2008). 'Building an innovation hub: a case study of the transformation of university roles in regional technological and economic development', *Research Policy*, **37** (8): 1188–204.

Zhang, F. and Gallagher, K.S. (2016). 'Innovation and technology transfer through global value chains: evidence from China's PV industry', *Energy Policy*, **94**: 191–203.

Zhang, J. (2009). 'The performance of university spin-offs: an exploratory analysis using venture capital data', *The Journal of Technology Transfer*, **34**: 255–85.

Zhang, S., Andrews-Speed, P. and Ji, M. (2014). 'The erratic path of the low-carbon transition in China: evolution of solar PV policy', *Energy Policy*, **67**: 903–12.

Zhang, S., Zhao, X., Andrews-Speed, P. and He, Y. (2013). 'The development trajectories of wind power and solar PV power in China: a comparison and policy recommendations', *Renewable and Sustainable Energy Reviews*, **26**: 322–31.

Zhang, W. and White, S. (2016). 'Global entrepreneurship and the origin of an ecosystem root firm: the case of China's solar PV industry', *Research Policy*, **45**: 604–17.

Zhang, X., Zhao, X., Smith, S., Xu, J. and Yu, X. (2012). 'Review of R&D progress and practical applications of the solar photovoltaic/thermal (PV/T) technologies', *Renewable and Sustainable Energy Reviews*, **16** (1): 599–617.

Zhao, Z.Y., Zhang, S.Y., Hubbard, B. and Yao, X. (2013). 'The emergence of the solar photovoltaic power industry in China', *Renewable and Sustainable Energy Reviews*, **21**: 229–36.

Zhi, Q., Sun, H., Li, Y., Xu, Y. and Su, J. (2014). 'China's solar photovoltaic policy: an analysis based on policy instruments', *Applied Energy*, **129**: 308–19.

Zucker, L.G. and Darby, M.R. (1996). 'Star scientists and institutional transformation: patterns of invention and innovation in the formation of the biotechnology industry', *Proceedings of the National Academy of Sciences*, **93** (23): 12709–16.

Zucker, L.G. and Darby, M.R. (1998). 'Entrepreneurs, star scientists, and biotechnology', *NBER Reporter*, p. 7.

Zucker, L.G. and Darby, M.R. (2007). 'Star scientists, innovation and regional and national immigration', NBER Working Paper No. 13547', National Bureau of Economic Research.

Zucker, L.G., Brewer, M.B., Darby, M.R. and Peng, Y. (1994). 'Collaboration structure and information dilemmas in biotechnology: organizational boundaries as trust production', *ISSR Working Papers*, **6** (2).

Zysman, J. (1994). 'How institutions create historically rooted trajectories of growth', *Industrial and Corporate Change*, **3** (1): 243–83.

Index